Changing School Mathematics

Changing School Mathematics

A Responsive Process

Edited by

JACK PRICE

Vista Unified School District
Vista, California

and

J. D. GAWRONSKI

San Diego County Department of Education
San Diego, California

American Association of School Administrators

Association for Supervision and Curriculum Development

National Council of Teachers of Mathematics

1981

Library of Congress Cataloging in Publication Data:

Main entry under title:

Changing school mathematics.

Bibliography: p.
1. Mathematics—Study and teaching—United
States.   I. Price, Jack.   II. Gawronski, J. D.
QA13.C44   510'.7'1073   81-11164
ISBN 0-87120-109-7 (Association for Supervision and
Curriculum Development)   AACR2
ISBN 0-87353-184-1 (National Council of Teachers of
Mathematics)

Printed in the United States of America

Contents

Part Two: Changing Mathematics Programs

Part Three: Changing, and Being Changed by, Others

Preface

As NCTM's *Agenda for Action: Recommendations for School Mathematics of the 1980s* took shape, it became evident that some help beyond its mere publication would be necessary to aid mathematics educators in implementing the recommendations. To meet this end, the Council's Publications Committee commissioned a professional reference book on change. The book was to be broadly based in the process of change and changing schools as well as in changing mathematics programs to reflect the agenda for the 1980s. *Changing School Mathematics: A Responsive Process* is the result of that initial charge of the Publications Committee.

The advisory committee for the book was selected to represent a broad range of mathematics educators as well as the full gamut of educational levels: school district personnel, a college professor, representatives from a state department of education and an intermediate unit, and a lay citizen. There were mathematics educators, a representative from the Association for Supervision and Curriculum Development, and the president of a private research and development organization. This group first met in August 1979 to brainstorm the book, develop the format, outline the chapters, and select possible authors. By December all the authors had been selected, and first drafts were received during the spring of 1980. The committee met again in September 1980 for the final editing of the first drafts and to see whether—as sometimes happens with committees—its "horse" had turned out to be a "giraffe."

The committee found that it is far easier to put together a diverse set of essays related only by topic than to develop a sequential book, asking each author to build on the unseen work of others. That the latter was done is a tribute to the authors and to the advisory committee, who were willing to spend their time reviewing and recommending. The final draft was sent to the NCTM Headquarters Office in February 1981 and placed in the capable hands of Charles Hucka and his staff, Charles Clements, June Carter, and Ann Butterfield. Now, by mysteries known only to them, it is finally in print.

To all of the above—the authors, the advisory committee, the NCTM staff, and our own staff, Kay Putnam and Kathleen Sudol—our heartfelt gratitude. If the final product reflects their work, the readers should be pleased.

Jack Price, *Editor*
J. D. Gawronski, *Coeditor*

Advisory Committee

J. D. Gawronski, San Diego County (Calif.) Department of Education
Eugene R. Howard, Colorado State Department of Education
Alan R. Osborne, Ohio State University
Jack Price, Vista (Calif.) Unified School District
Lynn Stuvé, Interface Network, San Diego, California
Ross Taylor, Minneapolis (Minn.) Public Schools

Introduction

*He that will not apply new remedies must
expect new evils; for time is the greatest
innovator.*
                    FRANCIS BACON

1

The "Agenda for Action" as a Potential Agent for Change in the Mathematics Curriculum

SHIRLEY A. HILL

F OREMOST among the five-year goals of the National Council of Teachers of Mathematics during the last half of the 1970s was the development of recommendations for the mathematics curriculum for the 1980s. That this goal was accorded top priority testifies to the determination of the Board of Directors that the Council would take a leadership role in effecting change.

Some of the assumptions underlying such a role are that (1) the curriculum should be a dynamic entity, constantly undergoing review and change; (2) an obligation of professional education organizations is to provide informed opinion and guidance on instructional and curricular issues; and (3) organized groups of professional teachers, with a consensus of dedication to improvement, *can* influence educational decision making at all levels.

Intent

The Council's recommendations were released in April 1980 in a booklet entitled *An Agenda for Action: Recommendations for School Mathematics of the 1980s* (see Appendix). The *Agenda*'s avowed intent is to effect positive change during the decade.

How realistic is this goal? As the first section of this book makes clear, change in the school curriculum is a complex process—difficult to analyze, more difficult yet to influence. No single group can manipulate all the myriad forces that affect and collectively determine educational policy and activity. Perhaps that fact is one of the school's greatest unheralded strengths as an institution. But another aspect of the same fact is that even though the school

3

on the one hand appears to favor the status quo and to be almost impervious to large-scale organized change, it is on the other hand particularly vulnerable to curricular fads that provide only superficial, short-lived results.

As the process of educational change is delineated in Part One, several prominent factors emerge. Two of them in particular might give reason to question the potential effectiveness of the recommendations. First, educational goals, and thus curricula, fall within the spheres of concern and self-interest of so many disparate groups. These groups exert differing and sometimes counteracting pressures on school policy. The pronouncements and positions of a single group will have little force without the support and reinforcing action of many other groups. Second, the most promising target for direct and planned change appears to be the individual school. Leadership in effecting change must exist at the school unit level, though it is not confined to that level. Can any set of recommendations by a national organization comprehend the diversity of need and interest at the local level?

To be sure, we do not lack for cynics who consider the intention of the *Agenda for Action* to be unrealistic and naive. But this view is to misread the ways in which the recommendations are intended to enter the process of change. Criticisms based on historical allusion to documents that, after a brief flare-up of professional interest, languish in the dust of library shelves are themselves based on a naive criterion. They assume that the influence of such documents resides in their being a set of directives either to be carried out or not. Thus their success or failure can be determined fairly quickly (within a few years) and holistically. Put simply, these documents say, Here is what the mathematics curriculum should be; now put it into practice.

It doesn't work that way, as the authors of Part One make abundantly clear and as our own recent experience in mathematics curriculum development ought to persuade us.

Perhaps the closest that mathematics educators came to the direct and holistic model for change was in the 1960s, when total mathematics curricula were written and put into practice in a limited number of places. These curricula, acknowledged as "experimental," had a short life. If one does not presently find such programs intact in the school curriculum, does this mean they were total failures, that they had no lasting influence? Of course not. The complexity of change makes it difficult to analyze, but one suspects that many of their undoubted influences were not strictly a part of planned change nor could they have been foreseen entirely by the developers. The failure of mathematics educators to make clear these subtle characteristics of curriculum dynamics accounts for the public's confused and erroneous perception of an "old math–new math" controversy.

The NCTM recommendations were developed with a conscious recognition and acknowledgment of all the aforementioned difficulties and with a clear-eyed determination to learn from historical experience rather than to repeat it. These goals, given the particulars of their intended functions and role and assuming their decade-long scope, are realistic. They assume the existing institutional structures and relationships and do not depend on a radical transformation of the schools or their social contexts. Primarily they demand shifts in priorities, attitudes, curricular focus, and public support within the bounds and constraints of present governing and administrative structures and modes of financing.

To clarify the intended role and uses of the *Agenda for Action*, it may help to proceed by negative example, to say what it is not:

- It is not a set of instructions for initiating or completing curricular change.

- It is not a set of guidelines for course content, organization, or the selection of instructional methods or materials.

- It is not a blueprint for a particular universal mathematics curriculum or syllabus.

What is it then? It is exactly what its title implies, an *agenda* for the decade. As an agenda, it sets out the things on which we need to focus attention. It proposes goals and suggests priorities and emphases. It identifies problems that need resolution and suggests directions for their resolution. It should be a focus for concerted action and a catalyst for local change.

The recommendations are a deliberate mix of the general and the specific. The general recommendations provide for considerable variation in specific implementation depending on local conditions. They are not rigid prescriptions but permit wide latitude in adaptation and in creative and imaginative interpretations. The items for more specific action are more often directed at particular groups who are seen to be instrumental in determining or implementing policy and process.

The audience for the recommendations is diverse: professional and layperson, educator and noneducator, mathematics specialist and the specialist from some other discipline. In short, there are recommendations addressed to every person who has a concerned interest in mathematics instruction.

Most significantly, the *Agenda* is a professional commitment—a promise from a large group of educators for systematic and focused efforts toward selected curricular goals. And it is a document that directs the eyes of the profession and the public to the *future*.

If the recommendations should do no more than accomplish in a serious way this last objective, they will have made a contribution. For the school, more than any other institution (except, I suppose, the church), must be oriented toward the distant future. Despite all the excited talk of "success" or "failure" of programs measured by the yardstick of year-end test results, the real test of an individual's or a generation's education will come, not this year or next, but years or decades hence. It is on the shoals of this truism that a narrow view of accountability, however laudable its intent, will founder and come apart.

Although such a truism appears obvious, recent years have nonetheless witnessed a decided tendency for public opinion to turn back toward past goals in education. The curriculum ought to maintain a sensitive balance: it needs the sound foundation and continuity of academic tradition, but at the same time it must try to anticipate future demands and adjust accordingly. The danger is that today's longing for the past is not so much an affirmation of basic principles and enduring values as it is a nostalgia for a simpler time of "remembered" educational success. (Such memories are largely a revisionist view of educational history.) The security of tradition is the last resort of a fear of the future.

Change in Mathematics Curricula

The factors just discussed apply generally to all school subject areas. Are there factors specific to mathematics that affect the process of change? Obviously there are. Many of the uniquely mathematical are discussed in the *Agenda for Action* as well as in Part Two of this book.

There is also the question of timing. Subjects such as science and mathematics, which are basic foundations of technology, are particularly sensitive to rapid technological change. In periods of drastic and dizzying change, the natural inertia of school curricula creates the real danger of a widening gap between education or training and social needs.

We are in such a period now. The present school mathematics curriculum reflects a time when it was presumed that all citizens needed some computational skill and perhaps a little algebra but only the few who would go into areas such as the physical sciences, engineering, and technical careers really needed more mathematics. Formal geometry was saved from being a "luxury" only by the traces of the belief that it somehow taught students to reason.

Today mathematics is in a period of significant expansion, not only in its uses and areas of application but in its own scope and definition. Most careers requiring postsecondary education, whether in college or technical institutions, demand some mathematical background. Mathematics is the principal problem-solving tool for our society. Mathematics pervades everyone's lives as workers, as consumers, and as citizens. It can no longer be questioned that a much broader quantitative and mathematical literacy is essential to all, and that mathematical skill beyond the computational and mechanistic is advantageous to an individual. It is advantageous both in economic terms and in the capacity to comprehend the complex interactions of many situations. It is advantageous to a society in maintaining or increasing its productivity.

The pervasiveness of computer technology is also related to the kinds of skills that are most advantageous. It expands and alters the very nature of the skills and knowledge that are most in demand. Computers and calculators compel significant revision in curriculum, and they also extend our horizons in instructional methodology.

School programs will either respond to the demands for change inherent in these developments or become irrelevant and obsolete. Two things can happen: change will be forced and thus haphazard, reactionary, and piecemeal; or it will be the consequence of systematic, planned anticipation of foreseeable needs. The purpose of the work to be done through the NCTM's *Agenda* is to encourage, aid, and provide guidance and focus for the latter.

Part Two of this book addresses the application of general issues and principles of change to the specific context of the mathematics curriculum.

Data Base

Recommendations for mathematics curriculum reform have rarely had the advantage of having an extensive foundation in recent, pertinent data. Timing again becomes a significant factor in that the latter half of the 1970s provided more information about classroom practice than we have ever had. In 1975, the National Advisory Committee on Mathematical Education (NACOME) issued its report of mathematics programs, K–12. It analyzed available data and made recommendations; but it decried the paucity of direct information about what goes on in the mathematics class-room, and its major recommendation was that this situation be rectified.

Partly as a result of this report, the National Science Foundation commissioned several status studies on classroom practice. At about the same time, the second round of mathematics assessment of the National Assessment of Educational Progress provided valuable assessment data. Thus by 1979, we had available considerable information on which to establish needs, priorities, and purposes. This background is another sense in which the *Agenda* is *realistic.*

But the agenda for the decade must also be responsible. As someone has said, "Education is everybody's business." All of society has a stake in its future as represented by the young. No matter how that ideal is interpreted, a responsible professional group should take into account the priorities of a broad segment of public opinion.

Toward this end, the NCTM, through the auspices of the Ohio State University and supported by the National Science Foundation, surveyed several different populations concerning their priorities and opinions about mathematics education. These data composed another part of the data base for the *Agenda.*

Such opinions do not, however, actually generate recommendations. The obligation of a professional group is to take them seriously into account and then to bring to bear professional knowledge and experience to determine what is needed or will be needed in future years. It is particularly important that professional recommendations be willing to go counter to the preponderance of public opinion if, in the best judgment of the professional group, this is required by either present or future need and is in the best interests of students. In such instances, a major thrust of professional leadership should be the effective education of the public. The public has a right to expect no less of professionals.

Acting on the Agenda

Part Three of this volume focuses on the implications of change in mathematics programs for this decade. The *Agenda for Action* makes clear that its purposes have not been served merely by its publication. It is an agenda; the action must follow throughout the decade and probably beyond. Many of the specific activities and conditions involved in acting on the recommendations are discussed in Part Three.

Given the premise that educational policy is largely determined at the local or sometimes the state level, the actions of NCTM should fall generally in the following categories: (1) public relations, (2) political action, (3) support and consultation for local and state professional efforts, (4) collecting and disseminating models or examples of individual, local, and state efforts that represent implementation of particular recommendations (the mathematicians' "existence proof," so to speak), and (5) producing guidelines, ideas, and instructional resources that aid in accomplishing the recommended goals.

The last two categories are the things organizations like the NCTM do best and are best set up to do. The third category recognizes an important function of published position papers like *An Agenda for Action.* It is not so much that they try to develop entirely new ideas, concepts, and plans for the profession to accept but that they represent a professional consensus that by its very statement can provide support and moral suasion for those who are trying to accomplish things educationally in their own locales. References to the *Agenda* already abound in state and local curriculum plans and political activities on behalf of school issues. These references are used to support creative and original efforts, even though the plans may not owe their genesis strictly to the NCTM agenda.

The first two categories, public relations and political action, are areas where organizations like the NCTM have perhaps been weakest. Yet it is unlikely that a professional group can influence educational policy and thus lasting curricular change without effective activity in both.

Political action must entail garnering support from many disparate groups, both within the profession and outside it, who have a direct interest in mathematics education. Statements and resolutions of support are only a beginning. The coordination of effort from different coalitions of groups or influential individuals toward the accomplishment of particular purposes is essential and is an ongoing function of organizational leadership.

The extent to which a public relations effort at the national level can be effective is an open question. As pointed out elsewhere in this volume, influence by citizens on policy decisions comes directly from only a small number of those who are most knowledgeable about issues and are most committed to action. Most of the public do not keep up with the details of educational debate, nor do they interact often with school policy makers or personnel. (Recent Gallup polls on public attitudes toward the public schools have shown a significant difference in opinion between those who have direct contact with public schools and those who do not.)

But neither policy makers nor influence wielders are themselves impervious to a broad and vague public opinion when it achieves some focus or manages to rally round a slogan. The decade of the 1970s provides some good illustrations. No individual or group set out deliberately to lead a movement called "back to the basics." Nor can the broad-brush portrait of the school as "a failure" be credited to any systematic planned publicity. Yet these instances of the pressures of perceived "public opinion" have had enormous influence on recent school programs and objectives.

Can this elusive public opinion be informed or affected by a professional organization's public information at the national level without the financing customarily used for public relations campaigns? It is no doubt difficult but probably not impossible.

The columnist George Will wrote, "The wax of the public mind is soft, but not so soft it cannot take an impression." It is no surprise that a member of that influential fraternity of editorial writers who have the advantage of the public's attention should make that statement. There is little doubt of the central role of the news media in forming public opinion. But in general the media have not shown notable ability or inclination to deal with the complexities and subtleties of the most profound educational issues. To rely on the media as the conduit for conveying the challenges to public support required for improvement in mathematics education will require tenacity, patience, and an unusual tolerance for frustration.

An organization with a large number of local and state affiliated groups can also use this network to carry out a coordinated concentration on local media contacts.

Whatever the acknowledged difficulties, the attempt to enlist broad public support should be a part of action on the agenda.

In publishing *An Agenda for Action*, the National Council of Teachers of Mathematics was not unaware of the fact that its plan was ambitious. It can be extremely discouraging to work for curriculum improvement for decades and then look back and wonder how much of the result was more apparent than real. A very thoughtful mathematics educator stated that "the basic tenet in the proposed instructional reorganization is to make arithmetic less a challenge to the pupil's memory and more a challenge to his intelligence." This proposal is perfectly consonant with several of the NCTM recommendations. But that sentence was written in 1935 by William A. Brownell.

It should not be supposed that the intention is to end this discussion on a discouraging note. The point is that we be realistic, that we recognize the remarkable difficulties in achieving change even when the purposes appear, on the surface, so eminently desirable. But we need to return once again to the crucial factor of timing. During the next few years the mathematics curriculum *must* respond to an immense challenge to meet changing social demands. The famous quote about the power of an idea whose time has come may have been overused lately, but it still expresses a profound concept. The *Agenda for Action* is an idea whose time is this decade. The decade of the 1980s is a decade for mathematics.

PART

Changing Schools

When old ways no longer produce the desired outcomes, the sensible recourse would be to experiment with new approaches. But individuals are apt at such times to cling even more desperately to the old and unproductive behavior patterns. They are dissatisfied with the situation; but the prospect of change arouses even more anxiety, so they seek somehow to find a road back to the old and (as they see it) more peaceful way of life.

GOODWIN WATSON
Concepts for Social Change

Any discussion about changing the mathematics curriculum must be deeply embedded in a discussion of the process of change and the process of changing schools. In Part One of this book a generalized development of the change process and strategies for implementing change are discussed. The material is based on the idea that the best unit for change is the individual school.

Williams and Cummings identify general efforts for education reform and their characteristics and effects. They also provide some evidence from research for successful or effective strategies for implementing school-based change as well as for identifying impediments to change. Miller describes the politics involved in the change process and the need to consider "attentive publics." Abilities to think in coalitional terms and to use a negotiated consensual process are described as important factors in the politics of change. Blanchard and Zigarmi review research related to the nature of schools and change in schools. They present a model for change and strategies that curriculum specialists and other educational leaders could use to diagnose staff attitudes and school settings for change as well as ideas for helping teachers change behaviors and maintain new behaviors. Lieberman and Miller share some effective strategies for changing schools by considering the individual teacher and the school as an organization. They elaborate on systems that support classroom change. These ideas are then illustrated in the case study of a particular project for change or improvement in an urban high school. Howard summarizes Part One in a chapter that describes the processes necessary for change in a school or school system that is responsive to the need for change. In detail he moves from needs assessment through action plans to implementation and evaluation.

The central theme of this entire section is that haphazard change has little chance for producing success. For improved curriculum to result, a systematic program of preparation, information, and support must be carried out. Effective change results from the cooperative efforts of all concerned.

A careful reading of Part One will set the stage for the remaining chapters, which deal specifically with how to change mathematics programs.

2

The Dilemma of American Educational Reform

RICHARD C. WILLIAMS
TERENCE R. CANNINGS

There are two kinds of fools: those who say "this is new and therefore better" and others who say "this is old and therefore good."

William Ralph Inge

C HANGE is part of life. A study of human history reveals that human existence has always depended on successful adaptation to the environment. An institution or society that does not meet new environmental challenges will eventually disappear. And this need to change is never ending. Alvin Toffler (1980) suggests, for example, that we in the modern Western world are entering what he calls the *third wave*—a new civilization with new occupations, life-styles, work ethics, and economic structures. If he is correct, the ramifications for future planning, institutional and societal, are astounding.

American public schools, in spite of popular opinion to the contrary, have over the years experienced important changes. The early colonial school system was itself an interesting blend of European traditions and the developing American culture. The mid–nineteenth century saw significant educational reform—the American common school movement. This was followed in the twentieth century by progressive era reforms and the more recent ferment of the 1960s and 1970s.

The purpose of this chapter is to provide a brief historical background of efforts for change in American education, to review related research on recent efforts toward educational innovation, and to present the dilemma facing those who wish to implement substantial educational reforms.

13

Brief Review of Attempts at
Change in American Education

Reform waves have washed against school beachheads for centuries. Some have carved new structures in the rock; many have simply returned to the sea without leaving any traces. The most notable nineteenth-century reform was the common school movement, which was the product of a widely based social movement. The common school, as advocated by Horace Mann and other reformers, was to enroll and serve children from all social classes and religions. The system was to be publicly governed and financed. Its purpose was to develop a common citizenry that was literate, numerate, and moral. It was a reaction against what was perceived as the dehumanization of industrialization, urbanization, and immigration.

According to Tyack, Kirst, and Hansot (1980, p. 256), "The creation of such a system was a reform of immense magnitude—indeed, the greatest institution-building success in American history." The various political parties were uncommonly similar in their educational reform programs (Tyack 1974). The movement resulted in establishing the one-room schoolhouses in rural communities and graded schools in towns and cities. As Wiebe (1969) and Howe (1976) suggest, the schools were quite similar in purpose, giving students a basic elementary education to enable them to enter the nation's political and economic mainstream.

At the beginning of the twentieth century the press for centralized control of schools gave educators the power to adapt the schools to what they considered to be the differing needs and destinies of students in a complex industrial society (Tyack, Kirst, and Hansot 1980). These "progressive" administrators became part of an internally hierarchical school system that was increasingly shielded from lay influence. They believed that schools should prepare students for specific societal roles and should sort and train young people according to the students' destinies and the nation's needs. This was an important shift from the earlier common schools, which emphasized the three Rs and civic morality.

These professionals developed numerous programs, and many of these programs, such as the creation of junior high schools and school counselors, remain a part of our present education system. Other reforms, such as team teaching and the core curriculum, have had a more checkered fate (Orlosky and Smith 1972).

A glance at reforms initiated since the launching of Sputnik in 1957 reveals a kaleidoscope of innovative programs. Some were designed to alleviate the problems of special groups, such as the disadvantaged, the gifted, and the bilingual. Others—minimum competency testing, alternative schools, and new math, for instance—were aimed at revising basic school practices. The federal government has played an increasingly heavy role in these educational reform efforts.

"Where Have All the Innovations Gone?" could be the title of a song on the educational chart of top hits. Although some significant changes have been implemented or partially implemented, a vast number of reform efforts either have failed or have not realized their expected impact. The most persistent changes have been additive, easily monitored, and supported by a new, and often strong, constituency. These changes have often provided additional human or material resources—for instance, teacher aides and supplemental reading materials. The reason so many other innovations have failed is that their design or implementation strategies have ignored the realities or the "culture of schools," as Sarason (1971) calls it. Often reform models have treated the school and its inhabitants as a neutral "black box" into which were poured ideas and resources and out of which would flow desired outcomes. School reformers have often misunderstood the organizational properties and realities of school systems.

The modern reform era has been increasingly marked by the clash between those seeking to maintain control and external reformers seeking to impose long-range and significant changes.

Often reformers have had to resort to litigation to achieve their goals. The courts have, for example, ordered some school boards to desegregate their schools, ordered state legislatures to equalize per capita spending among school districts, and ordered educators to alter their programs for handicapped children. State and federal legislatures have also mandated changes. For example, certain state legislatures have required schools to implement minimum competency testing, and federal legislation has often insisted on involving parents and citizens in decision making on federally funded programs. Federal regulations compel schools to eliminate sex discrimination in physical education and vocational classes.

All these moves are planned attempts by external agencies to rectify and improve the services schools provide. Levin (1976) has noted that many school reforms result from larger societal needs. Schools are being forced to change because their activities and outcomes are not viewed by some significant constituency as contributing to the larger social order. In Levin's view, the role of the schools is to carry out a responsibility delegated to them by the needs of the polity, that is, the state or organized community. Emanating from the polity is a set of demands or socialization objectives for transmitting the culture or reproducing and maintaining the economic, political, and social order.

It is perhaps too soon to determine if there is any basic structure of references to these demands for change. Goodlad (1978) suggests that American curricular and instructional reforms have a cyclical pattern in which various reforms appear, disappear, and then reappear in somewhat altered form. He believes the philosophical basis for change has alternated during the past thirty to thirty-five years between "hard and tough" and "soft and tender":

1. The mid-1940s to mid-1950s emphasized the more open, child-centered school—for example, Prescott's child study program.

2. The mid-1950s to early 1960s saw an emphasis on subjects and discipline-centered schools. The launching of Sputnik spurred a new wave of curriculum reform.

3. The 1960s to early 1970s brought a return to more open, child-centered schooling typified by the open classroom.

4. The early 1970s brought the "back to basics" movement, a return to the "hard and tough."

Although the emphases have changed and the reform demands have varied in intensity and objectives, have schools changed? Have cycles of reform, including the development of new vocabularies, resulted in any significant changes in instructional programs? *Living and Learning*, the stimulating Hall-Dennis Report (Hall, Dennis, et al. 1968) on Canadian education, suggests that "changes in education, no matter how sweeping, profound or ideal, are barren unless they bring about changes in the classroom" (p. 121). Goodlad and Klein (1974) report that in spite of the rhetoric, few innovations have found their way into classroom practice.

In summary, there have been several significant educational reform movements, each with discernible characteristics and effects. Recently the press for change has shifted somewhat from internal professionals to external critics and reformers. Some of the reforms have been lasting, and many of those have had common characteristics. Many reforms and innovations, however, either have failed or have not realized their intended purpose of altered and improved classroom instruction.

Brief Review of Research on Change

If we are to improve our ability to reform the schools through planned change, we should take a closer look at what we have learned from previous innovative efforts.

Researchers have increasingly turned their attention to the change process in education, and this has resulted in a literature that is growing both in size and comprehensiveness. We shall briefly summarize common findings from comprehensive and significant research studies on educational change. Two such studies are the I/D/E/A Research Division's Study of Educational Change and School Improvement—referred to here as the I/D/E/A Study (Bentzen et al. 1974; Goodlad 1975)—and the Rand Corporation's Federal Change Agent Study—referred to here as the Rand Study (Berman and McLaughlin 1973–78). (The purpose and design of these two studies are discussed in greater detail in chapter 4 of this volume.)

Some elements that emerged from the two studies that appear to be common to successful efforts for change are these:

- The local school site as the locus of change
- The need for local grass-roots "ownership" of the activity
- The stress on projects and activities that address real school-site problems
- The preference for practical materials and materials generated by teachers and in-service training activities

Projects that exhibit large-scale, district-wide, top-down strategies that ignore the needs and cultures of individual schools are not likely to succeed.

The consistency of the findings in these two major studies has been supported in other studies of the process of educational change. Another body of research studies and theoretical work, which does not directly pertain to the educational change process, also provides insights into why successful innovations are most likely to have the characteristics alluded to above. What is it about schools as a particular type of complex organization that makes them most responsive to innovations that display these characteristics?

An exceedingly important related research study is Lortie's *School Teacher* (1975), an in-depth examination of teachers—their training, selection, daily routine, values, expectations, and so forth. From Lortie's work emerges the picture of an occupation, essentially a craft, in which practitioners have largely learned their skills on their own, behind closed doors. Because these are self-learned practices, teachers value and cling to them to get themselves through the day. They welcome support from administrators and parents that will help them with self-identified, practical, day-to-day problems. They are not particularly enthusiastic about programs or activities generated on the outside that do not address immediate classroom needs. Top-down innovations generated in their own district also typically face an uninterested reception from teachers.

A second problem in implementing district-wide changes instead of changes at the local school site is the way in which school districts are organized. Weick (1976) has applied the concept of "loose coupling" in describing how school bureaucracies function. This concept draws on the imagery of a train that has several cars coupled to the engine. If the train is tightly coupled, the cars will be relatively responsive to movements of the engine. If it is loosely coupled, there is considerable slack between the engine and the cars, and therefore changes in the engine's direction are not felt in the cars until after considerable delay. Since in Weick's view school districts are loosely coupled, movements or initiatives from the top (the administration) to individual schools and, subsequently, to individual class-rooms will often be delayed or distorted as they go from level to level. What is more, the "ownership" of the innovations will be lost in the process, and the top-down innovation is likely to fail.

This lack of responsiveness on the part of teachers, caused by loose coupling, might be overcome if school administrators and instructional specialists would devote a substantial amount of time to making sure teachers understand, value, and feel a proprietary interest in the innovations that come from others. However, as we point out later, this kind of supervision is not likely to materialize.

Meyer (1977) differentiates between technical and institutional organizations and argues persuasively that the difference in orientation is critical to our understanding of how and why schools function the way they do. All organizations and occupations have a technical core, a set of practices that have been developed to solve occupational or organizational problems. There are considerable differences, however, regarding the certainty of those technologies. That is, some techniques or practices will definitely achieve the desired result, whereas others might not. In technical organizations, which include most competitive business enterprises and some public enterprises, managers give considerable attention to the technical core and take great pains to protect it from public scrutiny. That is, they do not want the public to break the organization's boundaries and become involved in decisions regarding this technical core. A lack of attention to the technical core may result either in the institution's demise or considerable public intrusion into its affairs, particularly if the institution's activities are viewed as threatening to the public welfare. Manufacturing defective tires or polluting the environment with toxic wastes are examples of such threats.

Institutional organizations—public schools, parks, and libraries—exist because the public intrinsically values them. They do not usually exist in a competitive environment, and they have not developed a fully defined and carefully researched technical core. Further, their administrators hesitate to point to specific, measured outcomes to justify their existence. They exist as social institutions. In schools, administrators more often use symbols instead of hard data to convince the public that schools should continue to receive its support. So, when administrators are asked if a school is good or not, the response typically is in such terms as the number of certified teachers, a published district curriculum, and an individualized institutional program. As often noted (e.g., Goodlad and Klein 1974; Lortie 1975), administrators give the technical core (instruction) far less attention than their administrative counterparts in more technical organizations. Given this institutionalized characteristic, there is little incentive for school districts to study educational innovations systematically and intensively.

But what if a district presses for district-initiated, top-down innovation and insists that teachers conform to the district's wishes? How do nonsympathetic teachers and administrators respond? The classic study of a similar phenomenon was completed by Gouldner (1954) in a gypsum factory. He found that employees, faced with rules and regulations that the organization expects them to follow, display different patterns of compliance and that compliance is related to the ways in which the rules were formulated and are enforced.

If the rules and regulations have been determined solely from the top or are being enforced in punitive ways ("You must conform or be penalized"), then there is great likelihood that the workers will only superficially implement or follow the rules or that they will circumvent or ignore them (mock bureaucracy). This would seem to be especially true with tenured teachers, over whom administrators have limited control. If, however, workers have been involved in determining the rules and regulations (representative bureaucracy), then true implementation is more likely to follow. See also Coch and French (1948) and Daft and Becker (1978).

Finally, Lawrence and Lorsch (1967), on the basis of their work and research on a number of different kinds of organizations, have formulated what is known as *contingency theory*. In part, this theory argues that there is no "one best organizational design" but that each organization's design and functioning is contingent on that organization's environmental and technical characteristics. Applied to schools, this concept would argue that there is no one best system—that school sites must adapt to their individual needs and situation and that district-wide initiatives toward change are not likely to be sufficiently flexible to meet local school-site needs. The locus of change, in short, must be the individual school.

In summary, the literature on educational change and other supporting research and theory suggest that districts will be most successful if their strategy for change includes the following features:

- Using the local school site as the locus of change
- Assuring that teachers or other users participate significantly in decisions at all stages in the change process, especially in those matters over which teachers have discretion
- Valuing innovations and activities generated by teachers that center on teachers' practical problems

The Dilemma

Having suggested the research and the theoretical base for an appropriate strategy for change in schools, we shall now discuss some school characteristics that militate against successfully implementing planned, long-term educational change. These give reason to consider future educational change a dilemma. We suggest five characteristics of schools and school districts that limit the ability of school districts to implement changes in the most desirable manner:

- Insufficient organizational slack
- Lack of incentives
- Scarcity of implementation skills
- Unstable organizations
- Technical uncertainty

Insufficient Organizational Slack

Changing a school's focus from its daily operating concerns to identifying problems, determining appropriate solutions, and implementing those solutions takes an enormous amount of time and energy. Often additional resources are required as well. Most schools and school districts, especially those troubled urban schools that most need this problem-solving strategy, simply do not have that kind of slack time or resources. The teacher's day is defined as meeting classes and teaching children, with an occasional faculty meeting after school. What is more, many districts are undergoing enrollment and financial cutbacks that greatly limit their discretionary time and resources. Some districts have employed "shortened days" to free up faculty to consider problems during the school day. We suspect that this practice will meet increasing community resistance as more and more families have both parents working, leaving no one to care for the children when they return early from school.

Lack of Incentives

The incentive system in most schools provides little reason for teachers to make an extra effort to improve the school's functioning. Teachers do not normally receive additional pay, status, or promotion for putting forth extra effort. This isn't to say that many teachers don't make such an effort for other reasons—such as recognition and the sheer joy of accomplishment—especially early in their career. But many learn that such efforts earn no rewards; there is no regular organizational mechanism to reinforce outstanding performance.

Scarcity of Implementation Skills

For a variety of reasons, both personal and organizational, most respected principals have developed a leadership style that does not often include involving teachers in organizational decision making. Instead, many principals have rather clearly marked out their responsibilities and strive to fulfill their tasks as efficiently as possible. Many, perhaps, have tried to include teachers in decision making but have not found teachers who are willing to be so engaged or have run into organizational barriers—lack of organizational slack, for instance. Thus, the system has developed leader-and-follower patterns that are fairly ingrained; few principals or teachers have the skills or incentives to change that pattern of relationships and responsibilities.

Yet, unless problem-solving skills are developed at all levels and teachers are more fully involved in decision making, it is unlikely that schools will be able to identify and implement innovations that are relevant to their school's situation.

Organizational Instability

Many schools, especially in urban settings, give a misleading impression of stability. That is, the building stands there and the organizational patterns persist. In reality, however, many schools are very transient institutions, with a great amount of turnover occurring within organizational and building perimeters. In some schools, especially in urban districts, annual turnover of teachers and principals is quite extensive. This tends to dilute or diminish the returns a district might realize on investments made in team building, staff development, and innovative practices.

The same can be said for school districts. Pincus and Williams (1979) have pointed out certain forces that almost inevitably return innovative school districts back into the mainstream of educational practice, such as the departure of leaders, changing political environment, and becoming captives of the benefits.

Technical Uncertainty

The technical core of schooling is comparatively weak. The best most educators can do is give rough probabilities about the degree of success that is likely in accomplishing certain goals. They can estimate the percentage of children, for example, who will become skilled in mathematics and reading, but in many districts there remains a core of students who simply do not benefit as hoped from the school's instructional practices. What is more, educators are unsure what to do about it! Put another way, the schools have persistent problems in fulfilling many of their primary goals, and the technical core for solving those problems is not at hand.

The implication of this technical uncertainty for educational innovation is that people *may* be willing to make the extra effort to improve the schooling system if they are sure their efforts will achieve the desired results. Without that assurance, many are unwilling to invest themselves often or continually in something that probably will not make a difference. Instead they retreat into their proved or at least workable practices, reluctant to get involved in innovative activities.

Conclusion

Why speak of a dilemma? Had this been written in the 1960s or early 1970s, we might well have written of the "promise" of educational change or "future" directions in educational change.

Essentially, our review of research has led us to feel that there are discernible strategies that are most likely to result in long-term instructional changes. The I/D/E/A and Rand studies and other supporting research suggest an approach that emphasizes (1) the local school site as the locus of change, (2) the importance of teacher involvement in innovation-related decision making, (3) the importance of the school's adoption of a problem-solving orientation, and (4) the need to focus on problems that teachers consider important to their day-to-day routine.

What seems equally clear is that the conditions do not exist in most public schools to implement such a strategy of change. Schools, especially many urban schools that are deeply troubled, simply do not have the time, resources, energy, stability, and leadership to engage in such processes for any extended period of time. In short, the dilemma is that although there is a clearer picture today about what to do, organizational and societal constraints severely limit our ability to actually do it.

By no means is this the time in our national educational history to throw up our hands and say nothing can be done, thereby letting schools decline into mediocrity. But neither is it the time to grasp for simplistic solutions to difficult problems. Evidence of this is everywhere; legislatures and courts and other policy-making bodies try to impose changes without paying adequate attention to the organizational realities of schools. Sustained and enlightened efforts are needed to implement long-lasting reforms. Changes can be made—changes must be made.

REFERENCES

Bentzen, Mary M., and Associates. *Changing Schools: The Magic Feather Principle*. New York: McGraw-Hill Book Co., 1974.

Berman, Paul, and Milbrey W. McLaughlin. *Federal Programs Supporting Educational Change*. Vols. 1–8. Santa Monica, Calif.: Rand Corp., 1973–78.

Coch, Lester, and John R. P. French. "Overcoming Resistance to Change." *Human Relations* 1 (1948): 512–32.

Daft, Richard L., and Selwyn W. Becker. *The Innovative Organization: Innovation Adoption in School Organizations*. New York: Elsevier, 1978.

Goodlad, John I. *The Dynamics of Educational Change*. New York: McGraw-Hill Book Co., 1975.

_____. "Educational Change: Where We've Been." Paper read at a conference of ASCD, Austin, Tex., 30 January 1978.

Goodlad, John I., and M. F. Klein. *Looking Behind the Classroom Door*. 2d ed. Worthington, Ohio: Charles A. Jones Publishing Co., 1974.

Gouldner, Alvin W. *Patterns of Industrial Bureaucracy: A Case Study of Modern Factory Administration*. New York: Free Press, 1954.

Hall, E. M., C. A. Dennis, et al. *Living and Learning*. Toronto: Ontario Department of Education, 1968.

Howe, Daniel W., ed. *Victorian America*. Philadelphia: University of Pennsylvania Press, 1976.

Lawrence, Paul R., and Jay W. Lorsch. *Organization and Environment: Managing Differentiation and Integration*. Cambridge, Mass.: Harvard Graduate School of Business Administration, 1967.

Levin, Henry M. "Educational Reform: Its Meaning?" In *The Limits of Educational Reform*, edited by Martin Carnoy and Henry M. Levin. New York: David McKay Co., 1976.

Lortie, Dan C. *School Teacher*. Chicago: University of Chicago Press, 1975.

Meyer, John W. "Research on School and District Organizations." Paper read at the Annual Conference of the Sociology of Education Association, San Diego, Calif., April 1977.

Pincus, John, and Richard C. Williams. "Planned Change in Urban School Districts." *Phi Delta Kappan* 60 (June 1979): 729–33.

Orlosky, Donald, and B. Othanel Smith. "Educational Change: Its Origins and Characteristics." *Phi Delta Kappan* 53 (March 1972): 412–14.

Sarason, Seymour B. *The Culture of the School and the Problem of Change*. Boston: Allyn & Bacon, 1971.

Toffler, Alvin. *The Third Wave*. New York: William Morrow & Co., 1980.

Tyack, David B. *The One Best System: A History of American Urban Education*. Cambridge, Mass.: Harvard University Press, 1974.

Tyack, David B., M. W. Kirst, and E. Hansot. "Educational Reform: Retrospect and Prospect." *Teachers College Record* 81(3) (Spring 1980): 253–69.

Weick, Karl E. "Educational Organizations as Loosely Coupled Systems." *Administrative Science Quarterly* 21 (1976): 1–19.

Wiebe, Robert. "The Social Functions of Public Education." *American Quarterly* 21 (Summer 1969): 147–50.

3

The Politics of Educational Change

JON D. MILLER

M OST efforts to achieve substantial educational change will require some involvement with what is usually called *school politics*. Many educational leaders have traditionally viewed school politics as only slightly more tolerable than a wave of the plague and have sought to avoid or minimize involvement with all things political. The purpose of this chapter is to argue that many of these negative views of school politics are based on a misunderstanding of the political process in local schools and that in fact the political process may even be supportive of, and helpful to, educational change.

In the analysis and discussion that follow, the focus will be on local schools and change-related issues in those schools. Some people may also be interested in macrolevel educational policies at the state or national level, but an interest in broader educational policies is a topic for another analysis and will not be treated here.

Who Cares about Local Schools and Educational Change?

The traditional view of school politics has suffered from two erroneous premises: (1) everyone has a set of attitudes toward local schools and local school programs, and (2) everyone ought to have a set of attitudes toward these issues.

The literature demonstrates conclusively that a substantial portion of the adult population has no attitudes on many topics, including local schools. (Hennessy [1972] has provided a useful distinction among opinions, attitudes, and belief systems. Opinions are lightly held reactions to nonsalient stimuli and are usually unorganized and subject to easy change. Attitudes are more organized and strongly held views of salient issues. Attitudes are usually stable and often the basis for such specific behaviors as voting or

24

contacting an official. Belief systems are totally structured and organized sets of attitudes—Marxism, liberalism, or conservatism, for example. For the purposes of this analysis, we are primarily interested in attitudes toward local schools, not simple opinions.) This is not to say that a person who has no interest in local schools has no interest in politics. A person may be vitally interested in agricultural policy, foreign policy, or other political issues and still not be interested in local schools.

This situation is not unique to local school politics. Group after group decries the lack of broad public interest in, say, agricultural economics, prison reform, international trade, environmental conservation, nutrition, welfare reform, or science. Yet the range of active issues in the political system far exceeds the scope of interest or the information-acquisition abilities of any individual. Unlike frontier ancestors who waited eagerly for old newspapers from the East, the modern citizen faces a virtual tidal wave of information daily, and it is the individual's lot to sort out that small segment of personal interest or value. The result is an inevitable process called *political and issue specialization* (Miller 1980).

Almond (1950) has suggested that persons who display a high level of interest in, and a high level of information about, an issue or cluster of issues may be conceptualized as an "attentive public" for that issue. First Almond's stratified model will be reviewed, and then that model will be used to suggest an approach to local educational change.

The Almond Model

In its simplest form, attentiveness refers to the selection by an individual of those activities in which one wishes to engage and those issues about which one wishes to be knowledgeable. We can distinguish three sets of decisions that individuals must make in the process of defining the scope and focus of their attentiveness: (1) whether or not to follow public affairs at all, (2) the issues or concerns on which to focus, and (3) the types, frequency, and intensity of overt political behaviors in which to engage.

To a large extent, political scientists and other students of public attitudes have tended to examine such overt behaviors as voting or campaigning as constituting political participation. The inclusion of the concept of attentiveness to issues is important because it helps define the information base on which actions might be taken and because it provides a measure of participation in nonelectoral politics. Rosenau (1974) has argued that many areas of issue are of sufficiently limited interest for them seldom to become major electoral issues but are decided by legislative and administrative processes.

The relatively low levels of public interest in, and knowledge about, local school issues is common to a number of other public-policy areas. For example, only a small proportion of the adult population reports a high level of interest in, or information about, foreign policy issues or regularly follows news events about relations among the nations of the world.

Recognizing the contradiction between the traditional view that all citizens should be knowledgeable about all major political issues and the actual distribution of interest and knowledgeability in foreign policy, Almond (1950) published a seminal analysis of the problem. He argued that the great majority of the adult population was generally uninterested in matters of foreign policy and developed an interest only in times of war and the activation of the military draft. Most of the time, the public monitoring of foreign policy is conducted by a much smaller segment of the population, which Almond termed the *attentive public*. This attentive public is generally better educated, has more income, and is more prestigiously employed than the general population (see fig. 3.1).

The attentive public is largely a self-selected group, and it communicates through specialized channels. In the example of foreign policy, these specialized channels might include the *Saturday Review, New Republic, Nation, National Review, Commonweal*, and *Foreign Affairs*. Many members of the attentive public tend to be members of such groups as the Foreign Relations Council or academic and professional organizations with strong interests in foreign affairs. Rosenau (1961, 1963) and others (Cohen 1973; Hero 1959, 1960; Miller and Semetko 1980; Mueller 1973) have operationalized the concept and provided a substantial empirical description of the attentive public for foreign policy. Miller, Suchner, and Voelker (1980) have operationalized the concept for organized science. Given the investment of resources necessary to become and remain knowledgeable in an area of issue, it is unlikely that any single individual would be a member of very many attentive publics.

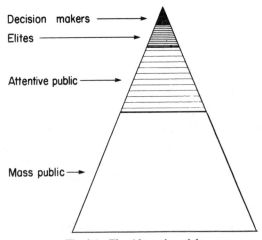

Fig. 3.1. The Almond model

The sources of specialized information on local school issues are available to most parents and teachers and are less difficult to gain access to than specialized information sources on many national political issues. Parent-teacher groups, teachers' professional associations, and unions all facilitate the flow of specific and specialized information about any given local school system. The availability of most schools for parent visitation and observation provides another important channel for acquiring information. But most of these specialized information channels are not open to nonparents and nonteachers; thus, it should not be surprising that few nonparents and nonteachers are attentive to local school issues.

Generally, the issues are formulated for an attentive public by a much smaller group of persons, called *elites*, who are speakers for various points of view. For national issues, these elites are usually corporate or labor leaders, academic leaders, or editors of some of the specialized communications media noted above. The elites debate among themselves and thus help to keep the attentive public informed about the problems and substantive policy alternatives. The elites also interact with the top of the pyramid, the decision makers. A number of studies have provided a rich literature on the elites in foreign policy (Cohen 1959; Hero 1960) and in a limited number of other policy areas.

For local school issues, the elites tend to be teachers and leaders of parent-teacher groups. It is important that teachers who are in fact providing leadership of this type see themselves in this role and understand its implications. In most local school systems, the relatively small number of attentives and the geographical and organizational unity of the system make it relatively easy for almost any individual or group of persons to inject an issue into the local system and to stimulate a discussion of it—especially once the focus is on attentives and not the total population.

At the top of Almond's pyramid of public opinion sit the *decision makers.* In foreign policy, the decision makers would include the leaders of the executive and legislative branches of the federal government and perhaps the leaders of selected corporations. Decision makers tend to be recruited almost exclusively from the elites in a given area. For those policy areas involving the federal government, there will be some overlapping of decision makers, but even here there is a relatively high degree of specialization (Fenno 1966; Orfield 1975; Price 1972). For local schools, the decision makers are the elected board and the principal administrative officers—the superintendent and selected assistants.

Almond (1950) argues convincingly that this pyramid of specialization—from the uninterested general public to the attentive public to the elites to the decision makers—does not undermine the basic tenets of a democratic society but rather reflects the differential ability of individuals in our society to devote the necessary resources and time to become and remain informed and active in a very specialized area. The public retains the final veto power; indeed, in times of war and military conscription the public interest has risen and has often forced the modification of a foreign policy. The role of American public opinion in ending the conflicts in Korea and Vietnam demonstrates this principle.

The Measurement of Attentiveness

In his original work, Almond (1950) suggested that to be considered attentive an individual must display a combination of both a high level of interest in an issue and a high level of substantive information about the issue, not simply a high level of one or the other.

Of the two dimensions, interest reflects the selective nature of the process. Since an individual cannot be attentive to all the possible issues or topics active at any point in time, it is imperative that one select those topics or issues of greatest interest in a relative sense. Viewed in this context, those persons who display a high level of interest in local school issues have already made an important decision about the relative place of education in their value systems.

The level of information about local schools is a somewhat more difficult construct, but important to the overall model. In many areas, for example, foreign or economic policy at the national level, the issues are framed in a technical language, and it is possible to develop simple measures of the extent of a respondent's knowledge in that specific area. With local school issues, the level of information needed is less technical and, as noted above, specialized information channels are available to parents and teachers. In this instance, the perception of adequate information may be more important than testable knowledge.

For local school issues, information may play a catalytic role. For example, some persons may have a high level of interest in school issues but doubt their own competence in the area. They would be less likely to interact with teachers, principals, administrators, or board members to promote their ideas, since persons who are uncertain of their knowledgeability on school issues are unlikely to join in disputes on most issues, especially on curricular issues. If, however, a person thinks of himself or herself as being competent in a topic, then that perceived knowledgeability may combine with a high level of interest to stimulate that individual to take a more active role in school issues.

To illustrate the importance of the joint occurrence of interest and information, think about some alternative combinations. Obviously a person with low interest and little information is most unlikely even to know that any school issues exist, to say nothing of having an attitude and taking an action. It is likely, however, that some individuals may have a high level of interest because of their role as parents or grandparents, but because of a lack of formal training or low self-esteem, they may think of themselves as incapable of making a meaningful contribution to discussions of school issues. The traditional image of the low-educated parent deferring to the professional authority of the teacher and principal illustrates this combination.

Similarly, it is likely that some people may have a reasonable level of knowledge but lack a sufficient level of interest to become active in school issues. For example, the parents of former pupils in a school system may be knowledgeable about school issues and may have the benefit of several years of experience but may take the position that they no longer have children in a given school system and that it is the business of those parents with children still in the system to decide current issues. Some persons may be knowledgeable about school issues as professionals (a trained educator, counselor, or related professional) but may be personally uninvolved and thus unwilling to make the effort to follow the specific issues in a local system. Some business or labor or religious leaders who could bring considerable knowledge or insight into local school issues may have placed a higher priority on other issues and be unable to devote the time or energy necessary to involve themselves in school issues.

When the different combinations are viewed in these terms, it is clear that attentiveness demands both a high level of interest in local school issues and a perception of adequate information concerning those issues.

Who Is Attentive to Local School Issues?

It is now appropriate to ask about the characteristics of that public attentive to local school issues. Fortunately, data from a recent national survey are available to provide an outline of this public. (To determine the size and composition of the public attentive to science policy, the National Science Foundation sponsored a national survey of adults under contract SRS–78–16839. The survey was based on a multistage cluster sample of households and produced 1635 completed interviews. The response rate was approximately 75 percent. The survey instrument was designed by Miller and Prewitt [1979], the data were collected by the Institute for Survey Research at Temple University, and the primary analysis of the data for the National Science Foundation was directed by Miller, Prewitt, and Pearson [1980].)

As a part of the 1979 survey, each respondent was given a card listing nine issue areas and was asked to indicate for each area if he or she was very interested, moderately interested, or not at all interested. (The nine issue areas were international and foreign policy, agriculture and farming, local schools, new scientific discoveries, economic and business conditions, minority rights, the use of new inventions and technologies, women's rights, and energy policy.) The survey results show that 38 percent of the adult population display a high level of interest in local school issues (see table 3.1). Women are significantly (at the .05 level) more likely than men to be very interested in local school issues, and interest in school issues peaks during the child-rearing years. The 1979 data show no significant associations between the level of interest and the respondent's level of formal education, occupation, or socioeconomic status.

After reporting their level of interest in each of the issues, respondents were asked to review the list again and classify themselves on each of the issue areas as very well informed, moderately well informed, or poorly informed. The 1979 survey data show a substantially lower self-assessment on information than on interest, with only 20 percent categorizing themselves as very well informed about local school issues (see table 3.1). As with interest, women tend to view themselves as knowledgeable on local school issues more often than men, and persons in the child-rearing years tend to be

TABLE 3.1

Percent of Adults Interested in, Informed about, and
Attentive to Local School Issues

	Very Interested	Very Well Informed	Attentive	N
All Adults	38%	20%	17%	1635
Sex				
Female	43	23	20	862
Male	32	16	13	773
Age				
17–24	26	14	11	309
25–34	40	21	18	361
35–44	55	25	23	258
45–54	54	34	28	234
55–64	27	12	9	228
65 and over	29	15	12	245
Education				
Less than high school	33	13	11	465
High school	42	22	19	550
Some college	39	20	17	382
Baccalaureate	40	28	25	146
Graduate degree	35	26	23	92
Occupation				
Prof. & tech.	34	22	18	258
Managerial	38	23	22	170
Sales	39	23	20	90
Clerical	46	24	21	283
Craftsman	38	19	14	168
Operative	39	18	15	235
Laborer	23	11	11	64
Farm worker	41	13	13	32
Service worker	41	16	13	187
Socioeconomic Status				
Low	36	16	14	339
Middle	41	20	17	728
High	39	26	24	298

more informed about school issues than persons at the ends of the age distribution. In contrast to interest, however, the perception of knowledgeability of school issues is positively and significantly (at the .05 level) associated with the respondent's level of formal education, occupation, and socioeconomic status.

Following the Almond model, those respondents who classified themselves both very interested in and very well informed about local school issues were classified as attentive to local school issues. The 1979 survey results indicate that 17 percent of the adults in the United States are attentive to local school issues (see table 3.1). Women are significantly (at the .05 level) more likely than men to be attentive to local school issues, and school attentiveness is highest during the child-rearing years.

The 1979 data suggest a class bias in participation in local school issues, with persons from high socioeconomic backgrounds being significantly (at the .05 level) more likely to be attentive to school issues than those from middle or lower backgrounds. From the preceding analysis, it is clear that this difference in attentiveness is based on class-differentiated perceptions of knowledgeability. Persons of low and middle socioeconomic status are not less interested in school issues but feel less competent to deal in those issues; thus they do not qualify as attentive in our terms. And it is not only a labeling difference, since the lower level of perceived competence almost certainly means a lower level of actual participation.

What Do Attentives Do?

Having profiled the attentive public for local school issues, we can now appropriately ask how attentives function and how their behavior differs from that of nonattentives. Unfortunately, almost all the literature concerning the behavior of attentives has focused on attentive publics for national issues (Almond 1950; Hero 1959, 1960; Miller, Prewitt, and Pearson 1980; Miller and Semetko 1980; Miller, Suchner, and Voelker 1980; Rosenau 1961, 1963, 1974). The only previous application of the attentiveness model to local issues is Miller's (1978, 1979) analysis of local planning issues.

It is possible, however, to combine the previous literature on local political participation (Banfield and Wilson 1963; Dahl 1961; Verba and Nie 1972) and the behavior of attentives at other levels to suggest the types of behavior that might be expected of those attentive to local school issues.

First, attentives are likely to focus on policy issues rather than individual cases or problems. This is not to suggest that their interest is entirely impersonal—most attentives to local school issues will be the parents of current or former pupils in the system; few nonparents or nonteachers will be attentive to local school issues. Much of the information acquired by local school attentives will come from direct contact as parents or teachers. Attentives would be likely to process this information and experience in policy terms, whereas nonattentives would be likely to seek a solution to a particular problem on an individualistic, or nonpolicy, basis.

In their study of political participation at all levels, Verba and Nie (1972) differentiate between individual, or "particularistic," contacting of officials (Where is my Social Security check?) and communal contacting of officials (urging a revision of the Social Security system). When this distinction is used, it is probable that local school attentives are more likely to engage in communal contacting than in particularistic contacting.

Second, attentives tend to work through informal channels and seek negotiated consensual solutions to issues. It is likely that persons who are attentive to local school issues will be active in parent-teacher or parent organizations; will know principals, superintendents, and some local school board members socially or professionally; and will be active in some local nonschool organizations and groups. This latter characteristic is particularly important if local school issues become overtly conflicting, since attentives are likely to be effective coalition builders.

Third, attentives can be expected to approach most local school issues with relatively clear conceptions of the problem and the solution. This is not to suggest that these views are hardened or nonnegotiable. Attentives tend toward negotiated settlements of differences, but they are not uninformed participants and they will not abandon their attitudes easily or without cogent reasoning. They are unlikely to defer to authority per se.

What Teachers Can Do to Facilitate Change

In the introduction to this chapter, it was suggested that teachers need not fear school politics and that the attentive public for local school issues can be an important ally in initiating, supporting, facilitating, or defending educational improvements in most school districts. It is now appropriate to return to the starting point and ask how an individual teacher or group of teachers interested in introducing educational changes into a system might be effective in this process.

First, it is important to recognize that teachers who are interested in improving a local school system or a curriculum within a school system are in fact themselves important members of the attentive public for local school issues. Too often, teachers have considered themselves outside the political process, although this is changing rapidly in most areas of the country. As members of the attentive public, these teachers will find that a little reflection will begin to identify a significant number of other persons who share their interest and who can be counted on to support reasonable improvements in local schools.

Second, it is essential to think in coalitional terms. Some educational changes can and should be accomplished routinely within the structure of the school system and will involve little external awareness, but most major changes will be noticed immediately by attentive parents. The teacher or group wishing to facilitate and support a given set of changes should see these interested parents as potential allies, not intruders into the system.

Too frequently classroom teachers are more comfortable dealing with a parent who wishes only an individual solution for a specific pupil and are unnerved by parents who suggest changes in policies or procedures. Many teachers and administrators tend to try to discourage these parents and will use a professional shield to suggest that the issues they have raised are not appropriate for discussion. This is a dangerous response. Parents who express policy-level concerns should be recognized as potential or actual school attentives and should be invited to learn more about the existing and proposed programs.

In this regard, it is important to be equally open and supportive of those attentives with either positive or negative attitudes. The appearance of providing special access for persons of one point of view may do long-term harm to a given program or proposal. Not all attentives will agree on most policy issues, but they are likely to be united on the rights of parents and interested citizens to inquire into school programs and to receive full and equal access.

Third, if an individual teacher or group of teachers wishes to initiate discussion on a set of programmatic changes that they are not able to implement without further assistance—either through authorization or through funding—it is possible to stimulate a debate on the issue by winning the interest and support of a segment of the public attentive to local school issues. The key is to select those persons who are already attentive to local school issues or are potentially attentive. Identify parents who display a high level of interest in various programs and invite them into your school or classroom to see what you are doing or to talk about what you want to do. It is important to recognize that the membership of an attentive public can be built as well as discovered. Given the relatively small number of persons who are school-issue attentives in most systems, the process of introducing an issue into the arena for discussion is relatively easy. Timing is important, however, and if active opposition is expected or encountered, the discussion and decision-making process should be lengthened to allow the maximum level of awareness to develop among attentives.

Fourth, if proposed educational changes will result in a major property-tax change, it should be recognized that this will activate a number of persons who are attentive to local tax issues who are not normally interested in school issues at all. Too often, local school leaders and attentives have viewed referenda on tax levies purely as a school issue and been surprised when large numbers of persons totally uninterested in school issues have organized to oppose these tax proposals. It is critically important to recognize that there is a separate attentive public for issues concerning local economic development (Miller 1978, 1979). When school issues demand increased tax support, it is important for school attentives, elites, and decision makers to begin the coalition-building process early and to seek to recognize and understand the attentive group for economic development. There are often important and logical grounds for a winning coalition, but a successful coalition for this purpose must almost always be initiated by school elites and decision makers.

Finally, the relatively small size of most local school political systems argues for the resolution of issues involving educational change through a negotiated consensual process rather than direct confrontation. Conflicts leave scars, and scars heal slowly in small political systems. If teachers and educational leaders correctly understand the system, a strong base for educational change and improvement can be built among the attentive public, and most issues can be handled without conflict. Whether a given issue is of sufficient importance to justify scars of overt conflict can only be judged by individual attentives and elites in the context of local circumstances, but the long-term price of open confrontation and conflict should not be taken lightly.

REFERENCES

Almond, Gabriel A. *The American People and Foreign Policy*. New York: Harcourt, Brace & Co., 1950.

Banfield, Edward C., and James Q. Wilson. *City Politics*. Cambridge, Mass.: Harvard University Press and M.I.T. Press, 1963.

Cohen, Bernard C. *The Influence of Non-Governmental Groups in Foreign Policy Making*. Boston: World Peace Foundation, 1959.

_____. *The Public's Impact on Foreign Policy*. Boston: Little, Brown & Co., 1973.

Dahl, Robert A. *Who Governs? Democracy and Power in an American City*. New Haven, Conn.: Yale University Press, 1961.

Fenno, Richard. *The Power of the Purse*. Boston: Little, Brown & Co., 1966.

Hennessy, Bernard. "A Headnote on the Existence and Study of Political Attitudes." In *Political Attitudes and Public Opinion*, edited by D. D. Nimmo and C. M. Bonjean. New York: David McKay Co., 1972.

Hero, Alfred O. *Americans in World Affairs*. Boston: World Peace Foundation, 1959.

_____. *Voluntary Organizations in World Affairs*. Boston: World Peace Foundation, 1960.

Miller, J. D. *Planning for the Future: A Summary of Citizen Views*. Final report submitted to FORGE, Inc. DeKalb, Ill.: Northern Illinois University Public Opinion Laboratory, 1978.

_____. "Political and Issue Specialization: A Behavioral Imperative." Paper presented to the Annual Meeting of the American Political Science Association, 28 August 1980, Washington, D.C.

————. "A Stratified Model of Attitudes toward the Politics of Community Planning." In *The Small City and Regional Community*, vol. 1, edited by R. P. Wolensky and E. J. Miller. Stevens Point, Wis.: University of Wisconsin—Stevens Point Foundation Press, 1979.

Miller, J. D., and Kenneth Prewitt. *The Measurement of the Attitudes of the U.S. Public toward Organized Science*. Report to the National Science Foundation in accordance with contract SRS–78–16839. Chicago: National Opinion Research Center, 1979.

Miller, J. D., Kenneth Prewitt, and R. Pearson. *The Attitudes of the U.S. Public toward Science and Technology*. Report to the National Science Foundation in accordance with contract SRS–78–16839. Chicago: National Opinion Research Center, 1980.

Miller, J. D., and H. Semetko. "The Attentive Public for Foreign Policy." Paper presented to the Annual Meeting of the Midwest Association for Public Opinion Research, 5 December 1980, Chicago.

Miller, J. D., R. W. Suchner, and A. V. Voelker. *Citizenship in an Age of Science*. New York: Pergamon Press, 1980.

Mueller, John E. *War, Presidents and Public Opinion*. New York: John Wiley & Sons, 1973.

Orfield, Gary. *Congressional Power: Congress and Social Change*. New York: Harcourt Brace Janovich, 1974.

Price, Don K. *Who Makes the Laws: Creativity and Power in Senate Committees*. New York: Schenkman Publishing Co., 1972.

Rosenau, James N. *Citizenship between Elections*. New York: Free Press, 1974.

————. *National Leadership and Foreign Policy: A Case Study in the Mobilization of Public Support*. Princeton, N.J.: Princeton University Press, 1963.

————. *Public Opinion and Foreign Policy*. New York: Random House, 1961.

Verba, Sidney, and Norman H. Nie. *Participation in America: Political Democracy and Social Equality*. New York: Harper & Row, 1972.

4

Models for Change in Schools

KENNETH H. BLANCHARD
PATRICIA ZIGARMI

T HE 1960s was a decade characterized by a tremendous influx of innovations in schools. Money was available and creative juices were flowing. But in the 1970s school districts and researchers alike began to realize that the innovative packages and projects they had introduced in the 1960s had not brought about expected changes. What one would often observe in a school where teachers had "implemented a change" would be a potpourri of practices, some more or less resembling what the developer of the innovation or the initiator of the change had in mind. Too often, teachers and curriculum consultants would agree that many of the innovations they were asked to implement were irrelevant and most of their efforts at change ineffective.

Researchers observed that many of the so-called innovations were, in effect, never implemented. On the basis of these shared observations it seems useful to try to increase our understanding of the process of change by looking at several "models of change." It is to be hoped that these models will improve our chances for success in implementing innovations in schools.

In this chapter we shall—

1. review what researchers have learned about the nature of schools and about change in schools;

2. discuss a general model of change based on Kurt Lewin's (1947) three processes of change—unfreezing (initial implementation), changing (adaptation), and refreezing (institutionalization);

3. share descriptions of several tools that curriculum consultants and other educational leaders can use in diagnosing the school setting and teachers' attitudes toward change.

Each of these sections will help curriculum leaders determine what they

36

can do to help teachers change their practices and maintain those new behaviors once the changes have been made.

What Researchers Have Learned about Change

Seymour Sarason (1971) and Matt Miles (1967) have probably done the most extensive work on the nature of schools and the imperviousness of many of them to change. Sarason observes that as one tries to change schools, one runs smack into the realization that there are no vehicles for discussion, communication, or observation that allow for alternatives to be raised and productively used for the purpose of change. He observes further that there are a number of programmatic and behavioral regularities in the ways that schools are run. Any challenge to these standard operating procedures is likely to engender more feeling than reason. Being more explicit about the characteristics of schools that make them difficult to change, Miles says that schools are—

1. subject to local public control;
2. nonvoluntary;
3. isolated from other socializing agencies;
4. characterized by vaguely stated multiple goals that are often in conflict with one another;
5. difficult to evaluate in a systematic way;
6. characterized by informal norms supporting autonomy and prohibiting interference;
7. characterized by a low degree of role differentiation, a low degree of interdependence, and a high degree of passivity.

In sum, the school "has little power to initiate, develop, grow, push things, or be disagreeable to anyone or anything" (p. 2). In many ways, these characteristics make schools intractable. When these organizational characteristics of schools are not recognized, change efforts are doomed to failure.

Recognizing the difficulty of effecting change in schools and the fact that few innovations are effectively implemented, Berman and McLaughlin (1975) undertook the *Rand Study of Federal Programs Supporting Educational Change* to determine what needed to be done for successful change to occur. The study identified four broad factors influencing the successful implementation and continuation of change efforts in schools: (1) institutional motivation, (2) implementation strategy, (3) institutional leadership, and (4) teacher characteristics. The criteria the study used to evaluate whether change had taken place were (1) the extent to which teachers changed, (2) the amount of student growth that took place, and (3) the continued use of new methods and materials once the project was completed.

Institutional Motivation

The first set of factors that the Rand Study found to be critical to project outcomes was labeled *institutional motivation*. The study observed (1975, vol. 3, p. 18) that

> a school district may undertake a special project and a school or teacher may agree to participate in the project, for very different reasons. A school may initiate a change agent project to address a high priority need, or it may start a project to ameliorate community pressures, to appear "up-to-date," or simply because the money is there. Similarly, teachers participate in a special project because they are "told to," or because it is their own idea, because of collegial pressure, or because they see the project as an opportunity for important professional growth. These different motivations significantly influence whether or not change is implemented and incorporated into day-to-day school practice.

Not surprisingly, the Rand Study found that teachers must be committed to an idea for it to be successfully implemented. In turn, the commitment of teachers is influenced by the motivations of administrators, by the planning strategies used in the project, and by the scope of the project. The study found that when teachers see that administrators aren't interested or supportive of a new idea, they feel that the personal costs involved in changing just aren't worth it. Secondly, teachers usually have little personal investment in project objectives or success when they have not participated in planning. Collaborative planning brought the consensus and support of all those who were involved. Finally, a third factor that influenced teacher motivation was the scope of change proposed by the project (1975, vol. 8, p. vii):

> The more effort required of project teachers and the greater overall change in teaching style attempted by the project, the higher proportion of committed teachers. In other words, complex and ambitious projects were more likely to elicit the enthusiasm of teachers than were routine and limited projects.

It was hypothesized that this was because ambitious projects appealed to a teacher's sense of professionalism and capitalized on teachers' intrinsic motivation to become better teachers.

Implementation Strategy

The second critical factor affecting the outcomes of change efforts was the project's *implementation strategy*. The Rand Study found that training should be *skill specific*. However, the researchers also observed that skill-specific training alone did not ensure long-term change in teachers or continuation of the project. They found that changes in behavior tended to be maintained only if (1) teachers received in-classroom assistance from credible resource people, (2) they had an opportunity to participate in decision making, and (3) regular staff meetings were held to solve problems.

These three activities helped teachers adapt new ideas to their individual classrooms.

Institutional Leadership

The third factor the Rand Study found important for successful implementation and continuation of a change project was *institutional leadership*. The study found that the support of central office personnel and, even more, the support of the principal are essential if successful change is to take place. Competent, enthusiastic leadership on the part of the principal and central office affected a project's implementation positively. Teachers who perceived that their principals valued changes tended to be more committed to implementing those changes in their classrooms. One way principals can demonstrate their support for change efforts is to become actively involved in project training. Finally, the active leadership of principals seemed to influence the quality of working relationships among teachers; this, in turn, enhanced the project's implementation and promoted the continuation of the project's methods and materials.

Characteristics of Teachers

The fourth set of factors that the Rand Study focused on was labeled *teacher characteristics*. These characteristics included the number of years of teaching experience, verbal ability, and "sense of efficacy." In general, the study found that the length of teaching experience was negatively correlated with all the outcome variables except teachers' continuation of project techniques. In other words, the more experience a teacher has, the less likely the project was to achieve its goal and the less likely the project was to improve student performance. "These relationships in large part are attributable to the fact that more experienced teachers are also less likely to change. . . . In effect they calcify with increased experience" (p. viii). The second characteristic, verbal ability, was significantly related to only one outcome measure—total improvement in student performance. The most powerful attribute in the Rand analysis was the teachers' "sense of efficacy"—a belief that the teacher can help even the most difficult or unmotivated students. This characteristic showed a strong positive relationship to all three project outcome measures.

In summary, the Rand Study found that intrinsic motivators are more effective than extrinsic motivators in securing teachers' actual involvement in an innovative project. The study also found a need for staff support activities—ongoing in-classroom assistance and regular staff meetings—and the need for the involvement of administrators in the change process. Finally, the study supported the notion that teachers with a high degree of confidence were most able to implement change.

The Process of Change

Whereas the Rand Study identified some specific factors that influence the successful implementation of change, Kurt Lewin (1947, pp. 5–41) developed a general model for understanding the process of change. Lewin said that there are three basic phases to a change process. The first phase is *unfreezing*, which prepares people in the organization for change. Next is the *changing* process, which takes place after people are ready to change and incorporates the steps that are taken to help make the innovation work in a particular setting. The final phase is *refreezing*—the systematic reinforcement and support of changes that have occurred during the change process.

Unfreezing

The aim of unfreezing is to motivate and make the individual or the group ready to change. It is a thawing out process through which the forces acting on individuals are rearranged so that they now see the need for change. In brief, unfreezing is the breaking down of the mores, customs, and traditions of individuals—the old ways of doing things—so that they are ready to accept new alternatives. For example, if a school administrator wanted to implement a new grading system, the unfreezing process might include hiring an expert to explain to the teachers the pros and cons of the new system and show them how it works. The goal would be to alleviate their fears about this grading system and to generate positive feelings about the change.

Changing

Once individuals have become motivated to change, they are ready to learn new patterns of behavior. This process is most likely to occur through two mechanisms: *identification* and *internalization*. Identification occurs when one or more models are provided, models from whom individuals can learn new behavior patterns by identifying with them and trying to become like them. Internalization occurs when individuals are placed in a situation where new behaviors are demanded of them if they are to operate successfully in that situation. Identification and internalization are not "either/or" courses of action; effective change is often the result of combining the two as a strategy for change.

In our example of the new grading system, the changing process might begin by having a few key teachers agree to pilot the new system for a semester and then report results to the rest of the teachers. If these results are good, the new system could be implemented with these key teachers (models) facilitating the process. For teachers now to be successful in their own eyes, they would need to use (internalize) the new grading system.

Force is sometimes discussed as another mechanism for inducing change.

Compliance occurs when an individual is forced to change by the direct manipulation of rewards and punishment by someone in a position of power. When the person in authority is present, behavior appears to have changed; however, when supervision is removed, it often reverts. Thus, rather than considering force as a mechanism of changing, we think of it as a tool for unfreezing.

Refreezing

The process by which the newly acquired behavior comes to be integrated as patterned behavior into the individual's personality or ongoing significant relationships is referred to as refreezing. As Schein contends (1968, p. 110), if the new behavior is internalized while it is being learned, this automatically facilitates refreezing because the behavior has been integrated into the individual's personality. If it has been learned through identification, it will persist only as long as the person's relationship with the original model persists (unless new surrogate models are found or social support and reinforcement is obtained).

This highlights how important it is for an individual engaged in a change process to be in an environment that is continually reinforcing the desired change. The effect of many innovative programs in schools has been short-lived when the people in charge of implementing the program do not reinforce new behavior patterns. In the initiation of a new grading system, the refreezing process might be accomplished by supportive visits by the principal with teachers during the grading period and continual recognition of their efforts.

Effective Change

For an effective change effort to take place in schools, all three of these phases must take place. There must be a period of unfreezing, a period when people are permitted to experiment with new behaviors, and finally a period of refreezing and follow-up assistance. The problem with most change efforts in schools is that the unfreezing and refreezing processes never take place. A new program is merely thrust on the school before anybody is ready or willing to engage in it. In the example of the new grading system, this would mean that the system is announced without much warning and then teachers are expected to implement it successfully without any support or assistance. After a while, it is likely that the teachers would become frustrated and return to the old grading system.

Diagnostic Tools

Determining How Much Unfreezing Needs to Be Done

Another concept developed by Kurt Lewin (1951) is very useful in determining how much unfreezing needs to take place in the school setting before an innovation can be effectively implemented. He calls this concept *force field analysis*. Lewin assumes that in any situation there are forces both driving for and restraining change. In terms of implementing an innovation in a school, examples of driving forces might be support from the central office and the principal, commitment of the teachers, and a sense of efficacy on the part of many teachers. Restraining forces are those acting to block the change—lack of administrative support and a commitment to old ways of doing things, for example. In essence, when forces that are pushing for change are counterbalanced by forces resisting change, there is no movement.

Before embarking on any attempt to implement an innovation, it seems appropriate for the persons responsible for that change effort to determine what they have going for them (driving forces) and what they have going against them (restraining forces). If innovators start implementing a change strategy without doing that kind of analysis, failure is inevitable. An example might be helpful.

In the spring, Joan Morris, an enthusiastic junior high school principal who had formerly been head of the mathematics department and a teacher in the school, decided to implement a new mathematics curriculum. She was committed to improving student achievement through humanistic educational techniques. In particular, she wanted to change the school's predominantly teacher-centered approach—in which the teachers always tell the students what to do, how to do it, when to do it, and where to do it—to a student-centered approach, in which students play a significant role in determining their own learning activities.

To implement the changes she wanted, Morris reviewed curriculum materials on her own, selected one set, and hired outside consultants to train teachers during the fall preservice workshop. She had hoped to participate in the training, but meetings at the central office kept her away. Once school started, she expected teachers to implement the new curriculum on their own.

Much to her chagrin, the principal learned late in the fall that few changes had actually occurred. In fact, none of the new materials were being used. She could not believe what had happened, particularly since the teachers had evaluated the fall workshop a "successful experience."

A month or two later, Morris was participating in a graduate seminar on the management of change. The class was discussing the usefulness of force field analysis, and Morris, who had shared her frustration with the class, was asked to think about the driving and restraining forces that had been present in her situation. In thinking about the driving forces, she was quick to name her own enthusiasm and commitment as well as pressure from the community to do something different in mathematics because of low achievement scores. However, she could not think of any other driving forces.

In thinking about restraining forces, Morris began to mention one thing after another. First of all, she said, she had never really had a good relationship with the present head of the mathematics department. This person had always felt that the principal was "too interested" in what went on in the mathematics department. In addition, some of the older teachers resented the principal's rise to her new position. Further, some teachers could see no advantages to the student-centered approach—it was against the way they had always done things. In fact, some felt they were doing a good job because a few honor students had received mathematics awards the previous year. Others resented the principal's forcing the change on their department.

In her own defense, Morris said there had been no way to get together with the whole department to talk about the need to change the curriculum before the fall workshop. Consequently, teachers were not involved in the planning. In addition, she realized that her own busy schedule had prevented her from participating in the training and providing follow-up assistance. In effect, when teachers encountered problems in implementing the new curriculum, they felt the principal did not understand their difficulties. Even experienced teachers who used the student-centered approach were not encouraged to help their colleagues. All this resulted in the majority of the teachers being a restraining force.

Some of the students (and their parents) also resisted the change because they were doing well under the old system. A number of the parents of students who were doing poorly in mathematics seemed to feel the need for more direction and supervision rather than less. They felt that the humanistic approach was too "soft."

As you can see, the restraining forces against changing the mathematics curriculum to a student-centered approach not only outnumbered but easily outweighed the driving forces and weighted the balance even more in the direction of a teacher-centered curriculum. Had Joan Morris done a force field analysis before beginning the change effort, she might have planned a more effective change strategy. In particular, she would undoubtedly have tried to engage in more unfreezing activities before charging on with the change effort.

Assessing the Attitudes of Teachers

Suppose this principal had really wanted to assess the teachers' attitudes toward the new mathematics curriculum and their readiness for change; what could she have done? She might have used a second diagnostic tool, the Concerns Based Adoption Model (CBAM) developed by Gene Hall and his colleagues at the Research and Development Center for Teacher Education at the University of Texas at Austin (Hall, Wallace, and Dossett 1973). The CBAM is composed of three parts: (1) stages of concern, (2) levels of use, and (3) innovation configuration. This model is based on several assumptions about change. The first is that change is a process, not an event. In other words, it takes time for unfreezing, changing, and refreezing to occur. Secondly, educational change is a personal experience. Teachers perceive a given innovation differently according to their personalities and past experiences. Finally, although the teachers' concerns vary in intensity and duration throughout a change effort, they usually diminish as the teachers become more familiar and skillful with the innovation.

Stages of concern. Individuals involved in a change effort generally progress through three global stages in their concerns about the innovation. When it is first introduced, self-concerns predominate ("How will this affect me?"). Once they begin to use the innovation, management concerns predominate ("Will I ever get it all organized?"). Only when these concerns are resolved do concerns about the impact of the innovation on learners predominate ("Are they learning what they need?").

Research on the CBAM has identified seven *stages of concern* (SoC) about an innovation. These stages have been verified; measurement procedures have been developed; and the framework has been extensively used in research and practice. These seven stages are described in figure 4.1.

Stages of Concern

SoC 0	Awareness	I am not concerned about the innovation.
SoC 1	Informational	I would like to know more about it.
SoC 2	Personal	How will using it affect me?
SoC 3	Management	I seem to be spending all my time getting material ready.
SoC 4	Consequence	How is my use affecting kids?
SoC 5	Collaboration	I am concerned about relating what I am doing with what other instructors are doing.
SoC 6	Refocusing	I have some ideas about something that would work even better.

Fig. 4.1

Levels of use. People also change over time in their use of new programs. Generally, as teachers become more familiar with an innovation they become more skilled and coordinated in its use and more sensitive to its effects on students. *Levels of use* (LoU) is a second dimension of the CBAM which describes changes in individuals in relation to their actual use of an innovation. Seven levels of use have been identified and verified (see fig. 4.2).

		Levels of Use
LoU 0	Nonuse	No action is being taken with respect to the innovation.
LoU 1	Orientation	The user is seeking information about the innovation.
LoU 2	Preparation	The user is preparing to use the innovation.
LoU 3	Mechanical Use	The user is using the innovation in a poorly coordinated manner and is making user-oriented changes.
LoU 4a	Routine Use	The user is making few or no changes and has an established pattern of use.
LoU 4b	Refinement	The user is making changes to increase outcomes.
LoU 5	Integration	The user is making deliberate efforts to coordinate with others in using the innovation.
LoU 6	Renewal	The user is seeking more effective alternatives to established use of the innovation.

Fig. 4.2

Innovation configuration. The third component of the CBAM is labeled *innovation configuration.* This concept allows us to see how much of an innovation is being implemented, that is, what parts of the innovation are being implemented, in what combinations. In other words, the developer of an innovation (the person or group who originally conceived it) has a model form in mind. By the time the developer's model is implemented, the model form may not be recognizable. At the very least, one or two components may have been adjusted to fit local needs. As a result, when the innovation is observed across classrooms, different patterns may be found as various components of the original innovation and various combinations are selected. How the component variations are selected, how they are organized, and the way they are used by teachers in different classrooms result in different operational forms of the innovation, or different *innovation configurations.* Again, a procedure for identifying configurations, a checklist, has been developed by the CBAM Project.

The Concerns Based Adoption Model is particularly helpful when it comes to diagnosing how much unfreezing needs to be done as well as identifying what kind of support teachers need as they are changing. Another example might be helpful.

Following a series of workshops in one large suburban school district, teachers were expected to implement a new science curriculum—one that had been developed and piloted by teachers. The workshops were planned on the basis of an ongoing diagnosis of teachers' concerns about the new curriculum (using a Stages of Concern questionnaire and a Level of Use interview). Initially, three short meetings were held to introduce teachers and administrators to the curriculum and to respond to their questions about when and how it would be implemented (SoC 1 and 2, LoU 1 and 2).

Once teachers began to use some of the science units (SoC 3, LoU 3), day-long in-service workshops were offered focusing on improving their management skills. Teachers learned more about how to locate and organize materials, how to group students, how to keep track of students' progress, how to schedule the units, and so forth. Later, when they began to feel comfortable using the innovation (LoU 4a) but wondered how they could assess its impact on students (SoC 4), training was provided in assessment and evaluation, questioning skills, and collaboration. Ongoing "comfort and caring" was provided by an in-classroom teacher advisor who responded to individual teachers' concerns and problems in implementing the innovation. Meanwhile, teachers were encouraged to make changes in the curriculum to fit their individual classroom and teaching styles. These changes were monitored with an innovation configuration checklist. When changes were made that were unacceptable from the science department's point of view, teachers received additional training or consultation.

Overall, the interventions that curriculum consultants and staff development specialists made in this change effort were planned to help minimize or resolve teachers' concerns with the curriculum, to help them use it more effectively, and to change aspects of the school district in which the curriculum was being implemented. What has been evolving from this project, and others that the Research and Development Center for Teacher Education is involved in, is a model of interventions. In this model or framework, facilitators are advised to plan and carry out interventions designed to do the following:

1. *Develop supportive organizational arrangements*—actions taken to develop policies, plan, manage, staff, fund, restructure roles, and provide space, materials, and resources to establish or maintain use of the innovation.

2. *Provide training*—action taken to develop positive attitudes, knowledge, and skills in role performance related to the use of the innovation through structured or preplanned activities.

3. *Provide consultation and reinforcement*—actions taken to encourage

use and to assist individuals in solving problems related to the use of the innovation.

4. *Monitor and evaluate the change effort*—actions taken to gather, analyze, or report data about the implementation and outcomes of a change effort.

5. *Communicate and disseminate information about the innovation*— actions taken to inform or gain the support of outside individuals or groups.

During the planning of these interventions, facilitators are advised to consider who is in the best position to intervene (the source of the action), with whom (the target), and under what conditions (using what media, in what location, with what degree of interaction between the source and target). These concepts are all in the developmental stage, but they may give curriculum leaders a way to think about and understand the alternatives open to them in facilitating a change effort.

Diagnosing Antecedents and Consequences for Change

Even when teachers understand an innovation and have a positive attitude toward it, this does not guarantee that they will be able to change their behavior in the classroom. A third diagnostic tool for facilitators to use is the ABC model, originally conceptualized by Fred Luthans and Robert Kreitner (1975). The model may be helpful in understanding the conditions supporting behavioral change. The A in the model stands for *antecedent*— anything that comes before the B. B stands for *behavior*, and C stands for *consequences*—anything that comes after behavior.

Antecedents are the conditions that have to be created before implementation takes place. Paul Hersey and Marshall Goldsmith (1980) have developed an ACHIEVE model that is helpful in determining what antecedents are needed before someone can effectively implement an innovation (see fig. 4.3).

Suppose innovators do a good job establishing the antecedents. All the things that are necessary for somebody to begin to engage in new behavior are set. Will that guarantee that the innovation will last? The answer is an emphatic no. Innovations have a chance of survival if the antecedents are taken care of, but the key to real long-term implementation and incorporation of an innovation is managing the consequences.

Three types of consequences can occur after somebody tries to use an innovation. There can be a positive consequence: the individual is rewarded, praised, paid attention to, and thus given some kind of positive pat on the back. When new behaviors are reinforced with positive consequences, the frequency of the new behaviors tends to increase.

A stands for *ability*. The first thing an innovator has to determine is whether the users of an innovation have the ability (knowledge and skill) to implement and use the innovation. If they do not, then the first thing that has to be provided is training.

C stands for *clarity*. Are users clear about the goals and objectives of the innovation and what is expected of them? Often innovations are sabotaged because no one knows why the innovation is being introduced or where they fit in relation to the change. Unless these points are clarified early in an intervention, the users' insecurities and concerns will make them resistant to change.

H stands for *help*. If an innovation is to be successfully implemented so that refreezing occurs, the innovator has to give users the kind of help they need. That help can include such resources as money, equipment, facilities, and appropriate supervision. Often innovations are forced on teachers without enough resources to make the program work.

I stands for *incentive*. People have to have some motivation or incentive for engaging in new behaviors. This means giving people an understanding of "what's in it for them" if they implement the innovation. Incentives are a very important condition to consider before an innovative program is introduced.

E stands for *environment*. Do the innovators really understand the uniquenesses required of the environmental setting for the innovation? Do they understand the customs, mores, and standard ways that people operate? Do they understand the external environment? Environmental considerations are a key to the effective implementation of any innovation.

V stands for *validity*. By validity, Hersey and Goldsmith mean the legal ramifications of engaging in a particular behavior. Are users sure about what they can and cannot do legally? In a school setting, validity would also involve whether people thought the innovation made sense. Has there been any research to prove that it will improve conditions in the school?

E stands for *evaluation*. How will the innovation be evaluated? How will the innovator determine whether or not the innovation is having the kind of impact he or she wanted? So often evaluations are done after the fact. An evaluation system has to be developed *before* an innovation takes place.

Fig. 4.3

Another type of consequence that can follow new behavior is negative. Individuals can be reprimanded or punished in some way. The tendency is that then the new behavior will decline. Negative consequences have no real place in the early stages of implementing an innovation unless they are used in some way to stop or diminish former behaviors.

Another "consequence" is no consequence at all. Behavioral scientists call that "putting the behavior on extinction." If innovators do not positively reinforce individuals for attempting behavioral change in the early stages of an innovation and, instead, ignore them or pay them little attention, then the frequency of the new behavior will diminish. This is what so often happens with change in schools. After an innovation is started, all the implementors go to a leave-alone, laissez faire style of leadership, expecting that the users will automatically engage in new behaviors. Once they find out after a period of time that very little change is taking place, they get angry and reenter the situation in a negative, punitive way. This results in the most frequent leadership style of educational innovators; we call it the "leave alone zap style." This style involves telling people what kind of changes to make, then leaving everybody alone, only to return when the new procedures are not being implemented and becoming very directive again. This style of implementation tends to create anger, resentment, and hostility in the users toward both the innovator and the innovation.

The key to the effective implementation of change is to "catch people doing something right" early in the intervention. As research has shown, if people are reinforced for almost any behavior in the desired direction, the frequency will increase. Users need to be reinforced at 100 percent frequency during the early stage of an innovation. After people have started to use the innovation, 100 percent reinforcement is no longer necessary. At that point more intermittent reinforcement works. The key here is that the frequency of the reinforcement schedule should start to diminish gradually over time until eventually users are comfortable with the change and are able to provide their own reinforcement. It is at this point that the innovation has moved through Lewin's three stages of change from unfreezing to changing to refreezing, and the change is well stabilized in the environment.

"There Is Nothing So Practical As a Good Theory"

In this chapter we have attempted to review what some researchers have learned about the nature of schools and to provide innovators with some tools to diagnose and implement change. We have presented these theoretical models with a firm belief in Kurt Lewin's classic statement, "There is nothing so practical as a good theory." In fact, we feel that good theory is merely "common sense organized." Thus, we have shared theoretical constructs to help curriculum leaders organize their thoughts and plans in such a way as to increase the probability that an innovation, once introduced, will actually be implemented, integrated, and maintained in the school setting.

Some innovators might be frustrated by this approach, since we have not provided a list of principles of change. Instead, we have described change *as a process* and provided models that seemed to simplify reality rather than to describe its complexity. An analogy might be helpful in understanding our approach.

In a speech at the University of Massachusetts, Herb Simon (1970) talked about the "information overload" that most people face. Most of the time we have too much information rather than too little. Often, in fact, this information overload seems to immobilize decision makers. Simon asked the question, "Suppose you were drowning in a sea of information? What would you do to save your life?" One approach is to drink up as much "water" as fast as possible in hopes that you will be able to lower its level enough to be able to stand. And yet reality says you would drown before that occurred. But why doesn't a fish drown when it's constantly swimming in a drowning environment? The answer is that a fish has a built-in monitoring system that helps it take from the water what it needs and not take what it doesn't need.

We hope the theoretical frameworks presented here will be that kind of monitoring system to innovators. We hope they will help curriculum leaders diagnose the conditions surrounding a change effort and therefore make better decisions about what actions to take. Clearly, people interested in change should realize that they do not need to know everything about change to implement an innovation, but at least they must develop effective strategies unique to their own environment for unfreezing, changing, and refreezing.

REFERENCES

Berman, Paul, and Milbrey McLaughlin. *Federal Programs Supporting Educational Change.* Santa Monica, Calif.: Rand Corp., 1975.

Connellan, Thomas K. *How to Improve Human Performance: Behaviorism in Business and Industry.* New York: Harper & Row, 1978.

Hall, G. E., R. D. Wallace, and W. A. Dossett. *A Developmental Conceptualization of the Adoption Process within Educational Institutions.* Austin: Research and Development Center for Teacher Education, The University of Texas, 1973.

Hersey, Paul, and Marshall Goldsmith. "A Situational Approach to Performance Planning." *Training and Development Journal,* November 1980, pp. 38–45.

Lewin, Kurt. *Field Theory in Social Science.* New York: Harper & Bros., 1951.

_____. "Frontiers in Group Dynamics: Concept, Method, and Reality in Social Science, Social Equilibria and Social Change." *Human Relations* 1 (June 1947): 5–41.

Luthans, Fred, and Robert Kreitner. *Organizational Behavior Modification.* Glenview, Ill.: Scott, Foresman & Co., 1975.

Miles, Matt. "Some Properties of Schools as Social Systems." In *Change in School Systems,* edited by Goodwin Watson, pp. 1–23. Washington, D.C.: National Training Laboratories, 1967.

Sarason, Seymour. *The Culture of the School and the Problem of Change.* Boston: Allyn & Bacon, 1971.

Schein, Edgar H. "Management Development as a Process of Influence." In *Behavioral Concepts in Management*, edited by David Hampton, p. 110. Belmont, Calif.: Dickinson Publishing Co., 1968. Reprinted in *Industrial Management Review* 2 (May 1961): 59–77.

Simon, Herb. Comments made at a symposium, School of Business Administration, University of Massachusetts, Amherst, fall 1970.

5

Supporting Classroom Change

ANN LIEBERMAN
LYNNE MILLER

T OO often efforts at reform have failed because they have viewed schools too simplistically and too mechanistically; they have been based on a view of schools and teachers as empty slates. Reformers should assume from the outset that it is teachers themselves who make change happen and make it endure.

If schools are to change, if routines are to be modified, if private solutions to complex problems are to be discussed in the open and challenged, then it is important for school improvers to follow some simple guidelines:

- Approach teachers as the real "experts" about teaching and learning. Assume that the personal style they have developed has some value and that helping teachers articulate what they are doing in their classes also helps them evaluate their work and thereby opens the possibility for improvement.

- Provide for rewards when they try something new. Rewards can be in the form of public recognition or private communication, but they are essential.

- Encourage dialogue among teachers; work to transform individual concerns into collective concerns. This can be done through meetings, informal get-togethers, and group projects during or after the school day.

- Realize that the power and influence of the principal is of paramount importance. The principal is responsible for the day-to-day operations of the school, for keeping the complex web of interaction under control, for presenting an image to the community. The principal is the critical person in making change happen (Lutz 1970).

52

Efforts for improvement also concern a larger, external environment. Schools are political institutions; as such, they are not completely autonomous (Tye 1980). Education is not neutral; it is a political enterprise.

All educational theories are political theories. All educational arguments and ideas contain value assumptions and include visions of utopias. Usually educational and social arguments are intertwined. What begins as an educational issue (selection, for example) is debated in social terms (selection is unjust because middle class children are selected and working class children are not). What is justified as an educational argument is really a social argument; social equality is the real point at issue. [Edetts 1973, p. 11]

The political nature of education has made school improvements difficult to initiate and implement. Educators are—most of us—quite dismissive of political intrigues and strategies; we see ourselves as above such concerns. And yet educators must act politically if positive improvements are to be made. We must learn what it is that we have to give away and what we can gain from political exchanges. We must develop a view of school improvements as a process of bargains to be made and remade; we must be willing to enter the political arena, and we must collectively and individually accumulate the skills of negotiation and survival that are essential in that arena (Ingram 1977; Elmore 1978).

In the 1970s knowledge about the process of school improvement increased a great deal because of the study of change in a variety of settings over time. Three major studies, alluded to in earlier chapters, have been helpful to people who work with, and think about, improving schools: (1) the Concerns Based Adoption Model (CBAM) studies of individual teacher development, (2) the Rand Change Agent study of federal improvement projects, and (3) the I/D/E/A study of school improvements in individual schools over a five-year period. Although each study was different in its focus and concerns, the three taken together have greatly enhanced the understanding of what it takes to improve schools.

Briefly, these studies indicate that the process of improvement happens simultaneously on two levels: the level of the individual teacher and the level of the school as an organization. On the individual level are the following understandings:

- Teachers adapt to change developmentally; "change takes time, is achieved in steps" (Hall and Loucks 1978).
- Teachers are motivated by a sense of personal efficacy, a belief that what they do in their classrooms makes a difference to the children they teach (Berman and McLaughlin 1980).
- Teachers want and need training in new ideas and techniques that not only is rich in information but also provides support for trying out the new techniques in their classrooms.

Thus any efforts for improvement in schools must begin with the concerns and needs of individual teachers; small steps toward improved practice are more important for the classroom teacher than any grand design. Further, it is important that teachers be actively engaged in the improvement process and that they see the connection between what they are trying to do and what effects those attempts have on students. Finally, teachers must be recognized for the things they are doing well already and supported by people and resources for the new behaviors and procedures they decide to attempt.

Although the individual commitment to improvement and change is necessary for all improvement projects, individual changes are not sufficient to ensure that schools will improve. Without an organizational commitment to, and engagement in, improvements, individual efforts by teachers in isolated classrooms do not hold much promise for sustained success. An understanding of the process of school improvement on an institutional level indicates the following:

- Schools, like individuals, adapt to improvements developmentally. Again, change takes time and is achieved in steps.

- Schools in which programmatic or school-wide concerns are linked to concerns of individual teachers are schools that have the greatest possibility for positive change.

- Schools must provide the necessary conditions for improvement, and motivation must come primarily from the principal.

Thus, on an organizational level, the small details and regular routines of daily practice must be the starting place for improvements. Further, positive change in any school requires the cooperative efforts of teachers as a group working together and of a principal who is capable of leading the effort for change.

The three major studies of change are useful in many ways to educators, though they certainly do not provide prescriptions for improvement. Rather, they give some perspectives to consider as each school and each teacher attempt in a unique way to make meaningful changes in schools.

Improving Schools: Some Promising Approaches

Some of the more promising approaches now emerging assure us that change in schools is indeed possible and can be successfully incorporated into the life of any given school. Among these approaches are *staff development, networking and dissemination,* and *problem-posing and problem-solving activities.*

Staff Development

Staff development can be viewed as either a remedial strategy to improve the teaching practices of individuals or a viable approach to school change. As a strategy for improving the school, staff development "considers the effect of the whole school [the staff] on the individual [the teacher] and the necessity of long-term growth possibilities" (Lieberman and Miller 1978, p. 1). In other words, staff development provides the opportunity for imparting professional knowledge to teachers and administrators as part of a school's general improvement program.

The approaches to staff development are varied, and different settings demand different strategies. The following is an example of a staff development approach that was built from the ground up, using research and theory to meet the needs of a specific situation. Miller and Wolf (1978) describe this approach with the representation in figure 5.1.

individual concerns

individual actions

dialogue about actions

collaborative action

organizational change and
support for change

Fig. 5.1

The strategy begins with working with individual teachers on their immediate classroom concerns (such as curriculum planning, testing, teaching techniques, or discipline) through seminars, classes, workshops, consultation, and in-class assistance. Teachers are encouraged to practice new behaviors and, after they have had some success in doing so, to share the effects of their actions with others. Discussion circles are instituted, and through them teachers can talk about issues of mutual concern and plan possible joint actions. Here, as before, such staff development activities as seminars, workshops, consultation, and classroom assistance provide structure and support for this kind of discussion and planning. As collaborative actions are tested, the school begins to change. An environment is thus created that supports individual change and incorporates organizational change. The process is cyclical; staff development can enter at any point.

This approach to staff development is an effective one for teachers, but it leaves out one important variable—administration. Staff development should include, as part of its overall plan, structures and opportunities for administrative development. At the point at which staff development encourages collaborative action and organizational change, the support and leadership of the principal is absolutely essential. Some districts have developed "executive academies" for administrative staff development; others have incorporated these activities into ongoing programs. In any event, one thing is clear: if staff development is to be effective as a strategy for improving schools, it must recognize the principal's central role in the process and must also provide opportunities for his or her growth and leadership.

Networking and Dissemination

The large-scale studies of school improvement in the last decade have revealed that it is necessary to provide continuing support for teachers and principals as they work on improving their schools. The process of change involves personal, organizational, and political interactions among a variety of people (Schiffer 1980). One structure that has been around for a long time but not used extensively in education is the *network*, an often informal and often temporary affiliation of organizations for a particular purpose.

There are times when the individual school is incapable of providing the kinds of support it needs (Parker 1977). Such a situation exists now in efforts at school improvement. Networks have some characteristics that differ significantly from those of a single institution; further, they are not difficult to form and can provide different sources and kinds of support. Parker describes five ingredients of networks:

- A sense of being an alternative to established systems—informal
- A feeling of shared purpose
- A mixture of information sharing *and* psychological support
- A person functioning as a facilitator
- Voluntary participation

Networks can be either formal or informal depending on a variety of factors. For example, the National Diffusion Network was formally created by the U.S. Office of Education to aid in disseminating curricular innovations. Its membership is tied together by meetings that are neither necessarily frequent nor geographically close. The overarching goal of disseminating ideas, along with a long history of providing practical help for the members, binds its people together. So, we have a formally constituted network with informal relations among the loosely connected membership.

The Teacher Center Network is another example of people in a similar role who get together to share in informal ways. Again, the constitution of the network was provided by government funds, but its activities are informal. And what keeps its people together is a sense of shared experience and shared problems.

Informal networks of principals, staff developers, and teachers are all possible and can provide sharing and support within an informal system that is outside the formal institutions to which most of us belong.

Networks, like other informal arrangements, make it possible to create temporary response mechanisms and to plan events, and they usually adapt to people's needs faster than the formal organizations. However, formal organizations can create informal means to make these systems come about. An example of this is the Metropolitan School Study Council (MSSC). This network of schools has been meeting since the late 1940s. It was taken over several years ago by a professor whose primary interests are in staff development. The network consists of a collaboration between a university and a group of school districts. The membership, then, comes from two formal institutions, but it is informal itself. Meetings are based on mutual interests, and different people come and participate. Membership is by district consent, but participation is voluntary. The substance of the meetings changes as the needs of the members change.

Another example of a network as an alternative system was the League of Cooperating Schools (Bentzen 1974; Culver and Hoban 1973; Goodlad 1975). From 1966 to 1971, under the sponsorship of the Institute for the Development of Educational Activity (I/D/E/A), a network of eighteen schools was formed in southern California. For a five-year period, people met, posed problems, struggled with curricular and organizational problems, and began to view the league as a group that provided innovative norms for its members. The group was informal. Members played both central and peripheral parts. For some, it was a powerful experience. Some attended regularly, and others participated only in an occasional meeting. This network stood as an alternative to the district or school group and served as a clearinghouse for many ideas and practices.

The focus of a network often helps bind people together and build commitment. In the league, this focus was improving schools; in the National Diffusion Network the focus is to spread curricular innovations. The important thing is that whatever the purpose of the network, people come to meetings already interested because *their* concerns are the concerns of the network.

All groups must have a task, a focus, a reason to be. But they must also provide for good interpersonal relations. This ingredient is especially powerful today, when people are being asked to do more with fewer supports. The very act of providing meetings and getting people together in an informal way on neutral ground can be a significant step toward engaging and supporting people and sharing resources.

Someone needs to be in charge of organizing the network, and this person needs to be someone who cares more about facilitating interaction and change than being recognized. This leader's major tasks are to persuade people to interact with each other and to share their knowledge and experience. He or she needs to know when to be directive (provide information) and when to be permissive (provide opportunities for the group). Building the group feeling is more important than building the leader's expertise. Thus a network can be a comfortable place for people who are accustomed to a competitive or tense environment.

It is essential that participation be voluntary. Participants attend network meetings because they are interested, because they enjoy the other members, and because there they can both contribute and gain support on a variety of levels.

A major goal of any school improvement is how to get new ideas into a school and have them understood, adapted, and used. A synthesis of several studies where the intent was to do just that reveals several useful generalizations (Emrick and Peterson 1978):

- Interpersonal influences in involving people in an improvement strategy appear to be critical.
- The style of both the interventionist and the staff and their capacity to work together is more important than substance or expertise.
- Local involvement and commitment appear to be most critical in getting the process going.
- Training that is significant involves face-to-face continuous personal involvement. "*Who* does it and *how* it is done is more important than what is done and when it is done."

When we put together what is known about networks and dissemination, we can add two additional principles to the list:

- People must be engaged on a personal level, involved in their own self-interest with people who are sensitive to them, their involvements, and their commitments.
- Structures can be created that do not need money or fancy buildings. They do need a focus, voluntary participation, informality, flexibility, and provisions for information and support.

Problem-posing and Problem-solving Activities

One difficulty in improving schools has been to get a massive problem in manageable proportion so that it can be described, worked on, and solved. Although researchers have made significant strides in understanding schools and classrooms, translating the findings into practical terms has remained problematic (Jackson 1968; Lortie 1975; Reutter et al. 1980). More than thirty years ago, action research (research done in the field with representatives from the research community working with teachers on their problems) made an important contribution to both research and practice (Corey 1953; Miel et al. 1952). At that time, as now, there was concern that teachers themselves be involved scientifically and systematically in problem-solving activities. Taba (1957) suggested that several questions might guide the identification of problems:

1. What conditions need to prevail both to allow and to invite teachers to state problems of importance to them?
2. In what context do problem identification and analysis produce the maximum of identification, allow for varied levels of involvement depending on the capacity and the insight of individuals, and permit significant problems to emerge?
3. What is the sequence of opening up a problem for research with teachers?
4. What is the timing factor? How is one to gauge how fast or how slowly to proceed and at which point to introduce which considerations?
5. What is a team pattern in guiding action research that yields the greatest possible combined competence? What is the role of the researcher? Supervisor? Teachers?

Knowing what the problems are remains only part of the task. How to engage people in attempted solutions brings additional problems. A contemporary strategy that considers and builds on action research offers us another approach to improving schools at the local level. This strategy attempts to fill the gap between research and practice through several key features (Tikunoff, Ward, and Griffin 1980):

1. A team is created consisting of teachers, a researcher, and a developer-trainer. Each member of the team has an equal voice and assumes equal responsibility.
2. The team's task is to find a problem, check with a like group to see that it has universality, gather evidence, and use the evidence as an intervention strategy.
3. The team make-up provides for both an insider's and an outsider's perspective.
4. The problem arises from the practical needs of teachers, and the evidence is collected in the reality of classrooms.
5. After the evidence is analyzed and teachers intervene in their own behalf, they can provide professional development for their peers.

In operation, people involved in this problem-solving approach developed some additional principles worthy of consideration:

- A team approach to problem solving at the local school level can involve both insiders' and outsiders' perspectives. Participants might all come from the same district but have different functions.

- Problems and the evidence collected to understand the problems better can serve not only as a strategy of staff involvement for the team but also as a source of school improvement for school and district.
- The collaboration of teachers, staff developers, and researchers working on a problem collectively helps fill the gap between research and development, theory and practice, and concept and reality.
- The kinds of questions or problems teachers pose are more likely to be ones that will gain commitment and the energy needed to work on improvement.

Problem-posing and problem-solving approaches do not have to be as elaborate and costly as the one described above. For example, in one high school it quickly became obvious that a major problem plaguing teachers was student truancy and tardiness. A committee of ten teachers and one administrator was formed to investigate the problem and to make recommendations for dealing with it. The committee first gathered data from each teacher about the extent of the truancy and tardiness. Results were plotted on a bar graph and distributed throughout the school. The data showed that although tardiness was a school-wide problem, truancy was not. Truant students tended to skip only one or two classes a day, usually not the whole day. Further, the evidence indicated that some hours of the day had higher rates of truancy than others. The entire school faculty was then polled about their reactions to the data and their suggestions for dealing more effectively with tardiness and truancy. After seven weeks of intensive work, the committee made a series of recommendations to the principal for *(a)* improving the accounting and reporting procedures, *(b)* identifying students who were the worst offenders, *(c)* initiating classroom guidelines for dealing with tardiness and truancy, and *(d)* developing a variety of programs and strategies for deterring these behaviors. The principal read the recommendations and then met with the committee. Together, the principal and the committee developed a multifaceted plan for the school. Thus, with no outside resources and in only seven weeks, a major school problem was systematically identified and researched and possible solutions were implemented.

Common Features of School Improvement Strategies

Although staff development, networks and dissemination, and problem-posing and problem-solving approaches are markedly different in content and focus, they do share some common features associated with successful school improvement projects. All three approaches (1) are grounded in notions of *linkage* and *developmentalism,* (2) have an operating style that can best be called *systematic ad hocism,* and (3) allow specifically for *local adaptation.*

Linkage means simply "linking two organizations or bringing information from one place to another" (Lieberman 1977). Linkages can take place across or within organizations. In either instance, the role of the linking agent is central. She or he functions both as an observer of an organization and as a participant in its improvement; she or he must move freely from one environment to another and try to match needs with resources, bringing together the most workable parts of separate systems. Linkage requires commitment to the process of change in a school, and it requires that the linking agent always be a present and active participant in the change. Such a role can be assumed by many people already in school districts (curriculum specialists, principals, teachers).

Developmentalism refers to an understanding of how systems and people change. The developmental approach is a gradual one: it begins where people are at the present and provides a variety of structures and opportunities for future growth and movement. Such structures and opportunities are designed to challenge what is "given" in an environment without undermining the strengths that people bring to their work. A developmental perspective views people and institutions as having the potential for change and provides them with the tools and supports to undertake that change in a positive way (Miller and Wolf 1978).

Systematic ad hocism refers to a style of working in a change project in schools which is characterized neither by rationalized planning nor by "seat of the pants" management. It is based on four premises:

- It is more important to have a map than an itinerary. A map provides conceptualizations of where one wants to go without looking at only one way to get there.
- It is important to have long-range perspectives in planning. Long-range perspectives allow for a number of alternatives from any given action without limiting the line of vision to one and only one desirable end.
- It is important to be adaptive. Responsiveness to an organization means being active and challenging, and it allows for change both in the organization and in the change itself.
- It is important to have a clear set of underlying principles and values about what needs changing in schools and how those changes should happen; a strong value base is the "system" that guides the seemingly ad hoc activities of change.

Finally, *local adaptation* means that each setting approaches improvements in a different way, that general plans are altered to accommodate specific situations, and that often it is necessary to reinvent the wheel. Local adaptation also means that people most affected by an improvement effort are those most involved in it, that they develop a commitment to something new by having the opportunity to shape it and make it their own.

Although the notions of *linkage, developmentalism, systematic ad hocism*, and *local adaptation* may seem general and abstract, they can be understood in concrete terms when examined in the context of a successful school improvement project involving the mathematics faculty of an urban high school.

Case Study of an Improvement Project

A staff development project for school improvement had been in operation for an academic year at an urban high school of moderate size. During that year, individual teachers had participated in coursework, workshops, task forces, seminars, and classroom assistance activities. Throughout the year, groups of teachers had come together temporarily to deal with issues of shared concern—how to deal with student truancy, how to evaluate student progress, and how to modify the use of the school's intercom system, for example. These groups were facilitated by staff developers who, though from outside the school system, had extensive school experience and access to human and material resources (linkage).

One group that formed that first year was composed of four mathematics teachers who were encountering similar problems in teaching their freshman classes. They were troubled by the varying ability levels of their students (all of whom were new to the school) and the difficulty in dealing effectively with so many levels in one forty-two-minute period a day. The group decided to work together during the summer in a staff development course; their goal was to decide on a workable solution to their common problem. By the end of the summer, the group had not managed to solve their problem, but they did manage to decide on a new approach. Through the staff developer, the mathematics group contacted a consultant who had developed a curriculum for dealing with multilevel mathematics groupings (more linkage).

When the new semester began in the fall, the group worked with the consultant in weekly meetings. By Christmas vacation, they decided once again to revise their plan. Though helpful in many ways, the consultant seemed more intent on having the teachers adopt the program he had developed than on assisting them in developing their own solutions to their own situations. The four teachers thought that many of the instructional strategies suggested by the consultant were useful and could be applied to their school's situation; they also felt that adaptations and modifications were essential. When school reopened in January, the consultant did not return, but the teachers continued to meet to plan improvements for the teaching of freshman mathematics (local adaptation).

From January until spring vacation, the teachers developed an improvement plan. They decided that by organizing the school schedule so that all freshman mathematics classes would meet at the same time, they could pool their resources. For the first six weeks of the term, students would meet in their originally assigned classes for diagnostic activities. At the end of that period, students would be reassigned according to their ability levels. Every six weeks thereafter, they would be reevaluated and reassigned. And every six weeks, teachers would rotate among the various classes.

With their plan in hand, the four mathematics teachers began discussions with their department chairperson, who presented the plan to the entire department at a regularly scheduled meeting. After some discussion, the department approved the plan. At the end of April, the chairperson, the original group of four mathematics teachers, and the staff developer met with the principal to discuss the plan. The principal had been advised all along that the mathematics department was involved in developing a new approach to freshman instruction; he was very supportive of the proposal and called in the scheduling staff to discuss the nuts and bolts of the intended reorganization. When school reopened in September, the new freshman mathematics program was in operation. Throughout the year, the teachers continued to meet and to make modifications in their original plan. The entire process took two calendar years and went through a variety of stages and transformations (developmentalism).

The role of the staff developer in this entire process was essential. She provided opportunities for teachers to think about their own teaching, and she organized activities that promoted discussion among colleagues. She was attentive to the expressed needs of individuals, and she was able to bring groups of people together who shared similar concerns. She allowed the group that formed to take the lead in dealing with the issues that concerned it, and she acted as a resource-gatherer and group facilitator throughout the two years of the project. She reacted to needs, concerns, and interests as they were expressed. She also acted from a clear set of values and assumptions about individual professional growth and school improvements. She helped teachers develop a system for dealing with emergent problems (systematic ad hocism).

This case illustrates the use of these concepts in action. The concepts can be useful in thinking about one's plans and in anticipating some of the activities. They also are useful in understanding that movement in several directions at the same time or two steps forward and one step backward is not only a legitimate way to work but may, in fact, be the way schools improve.

REFERENCES

Bentzen, Mary M., and Associates. *Changing Schools: The Magic Feather Principle*. New York: McGraw-Hill Book Co., 1974.

Berman, Paul, and Milbrey McLaughlin. "Factors Affecting the Process of Change." In *Schools, Conflict and Change*, edited by Mike M. Milstein. New York: Teachers College Press, 1980.

Corey, Stephen. *Action Research to Improve School Practice*. New York: Bureau of Publications, Teachers College, 1953.

Culver, Carmen M., and Gary J. Hoban. *The Power to Change: Issues for the Innovative Educator*. New York: McGraw-Hill Book Co., 1973.

Edetts, J. *The Sociology of Educational Ideas*. London: Routledge & Kegan Paul, 1973.

Elmore, Richard. "Organizational Models of Social Program Implementation." *Public Policy* 26 (2) (Spring 1978): 185–228.

Emrick, John, and S. Peterson. *A Synthesis of Findings across Five Recent Studies of Educational Dissemination and Change*. San Francisco: Far West Laboratory, 1978.

Goodlad, John I. *The Dynamics of Educational Change*. New York: McGraw-Hill Book Co., 1975.

Hall, Gene, and Susan Loucks. "Teacher Concerns as a Basis for Facilitating and Personalizing Staff Development." *Teachers College Record* 80 (1) (September 1978): 36–53.

Ingram, Helen. "Policy Implementation through Bargaining: The Case of Federal Grants-in-Aid." *Public Policy* 25 (4) (Fall 1977): 499–526.

Jackson, Philip. *Life in Classrooms*. New York: Holt, Rinehart & Winston, 1968.

Lieberman, Ann. "Linking Processes in Educational Change." In *Linking Processes in Educational Improvement*, edited by Nicholas Nash and Jack Culbertson. Columbus, Ohio: University Council for Educational Administration, 1977.

Lieberman, Ann, and Lynne Miller. "The Social Realities of Teaching." In *Staff Development: New Demands, New Realities, New Perspectives*, edited by Ann Lieberman and Lynne Miller. New York: Teachers College Press, 1978.

Lortie, Dan C. *School Teacher*. Chicago: University of Chicago Press, 1975.

Lutz, Frank. *Toward Improved Urban Education*. Worthington, Ohio: Charles A. Jones Publishing Co., 1970.

Miel, Alice, and Associates. *Cooperative Procedures in Learning*. New York: Bureau of Publications, Teachers College, Columbia University, 1952.

Miller, Lynne, and T. Wolf. "Staff Development for School Change: Theory and Practice." In *Staff Development: New Demands, New Realities, New Perspectives*, edited by Ann Lieberman and Lynne Miller. New York: Teachers College Press, 1978.

Parker, L. Allen. *Networks for Innovation and Problem Solving and Their Use for Improving Education: A Working Paper*. Cambridge, Mass.: Center on Technology and Society, 1977.

Reutter, Michael, Barbara Maughan, Peter Mortimore, and Janet Ouston. *Fifteen Thousand Hours*. Cambridge, Mass.: Harvard University Press, 1980.

Schiffer, Judith A. *School Renewal through Staff Development*. New York: Teachers College Press, 1980.

Taba, Hilda. "Problem Identification." In *Research for Curriculum Improvement*, edited by Arthur W. Foshay and James A. Hall. Washington, D.C.: Association for Supervision and Curriculum Development, 1957.

Tikunoff, William, Beatrice A. Ward, and Gary Griffin. *Interactive Research and Development on Teaching: Final Report*. San Francisco: Far West Laboratory, 1980.

Tye, Kenneth A. "Politics and Organizational Development." In *Schools, Conflict and Change*, edited by Mike M. Milstein. New York: Teachers College Press, 1980.

6

The Responsive School System: An Organizational Structure for Systematic Change

EUGENE R. HOWARD

T HE authors of previous chapters have given a historical overview of educational change, have described several theories and strategies for change, and have, in general, summarized the most advanced thinking about how schools can be improved.

This chapter proposes that schools and school systems need substantial reorganization so that the promising theories and strategies defined in previous chapters can operate in a facilitative, nurturing setting. This chapter also suggests that in a period of rapid technological, economic, and social change our schools must become more responsive to the "attentive publics" described in chapter 3. Such attentive publics include parents and other members of the adult community who have a stake in the quality of the schools as well as teachers and students, those groups most immediately affected by any innovation.

An organizational structure to facilitate this responsiveness will be defined sufficiently well for it to be built within any school district or individual school. In a sense, this chapter will provide district and individual school leaders with an organizational blueprint for facilitating systematic change.

The Responsive School System—a Rationale

In oppressive systems, people serve the needs of the institution. In responsive systems, the institution serves the needs of the people. School systems viewed as oppressive are destined to continue to lose public confidence and support to the extent that they fail to listen and respond actively to the voices of their attentive publics.

Most school systems and individual schools, however, are not organized to be responsive. School board members as individuals often attempt to carry messages from their constituencies to the board as a whole or to the superintendent. They turn their home telephones and family rooms into miniature complaint bureaus in a futile attempt to determine what school improvements the public wants and will support. Such attempts are futile because they are not sufficiently comprehensive. Too many voices are not heard and too many questions are not asked. Controversy takes on the disguise of consensus.

Organizing a responsive school system is extremely difficult because if school leaders listen to large segments of the attentive public, they will hear voices calling for different and, to a large extent, mutually exclusive priorities for educational change. If educational leaders are responsive to one set of demands, they run the risk of alienating the proponents of a different set. There does not exist in this country anything close to a public consensus regarding what constitutes an educational program of high quality. In the absence of such a consensus, responsiveness becomes a politically delicate process.

The alternative to responsiveness, oppressiveness, does not offer school leaders an easier path. The authors of chapter 2 have described how autocratic, imposed, top-down processes of change contain the seeds of their own destruction. Teachers tend to respond to oppressiveness by protecting themselves through unionization, subtle subversion, and downright sabotage. In oppressive organizations some people burn out quickly; they cannot affect the conditions that control them. Pupils in such schools tend to react with apathy, unwilling conformity, withdrawal, and rebellion. School board members who behave in an arbitrary and capricious manner without checking with their constituents run the risk of becoming ex–school board members.

Much of the vitality of our public schools has come from the system's responsiveness. The attentive public feels a sense of pride and ownership when its aspirations for schooling are translated into instructional programs. The history of American schooling is to a large extent a history of change in response to public pressure. Schools have not always offered, for example, driver education, vocational education, career education, special education (fifty-seven varieties), counseling programs, drug and alcohol abuse programs, sex education, conservation education, bilingual education, multicultural education, or environmental education. Such programs are now offered because of public demands that they be offered.

School leaders cannot ignore public demands without weakening local control. Pressure groups have learned that if they are ignored at the local level, they may be heard by state legislators, the Congress, or the courts. We now have federally imposed and state-imposed regulations on bilingual education, English as a second language, the education of the handicapped, and the education of pupils from low-income families. The courts have mandated desegregation and integration, due process for suspensions and expulsions, and the elimination of discrimination on the basis of sex. Equal educational opportunity, an elusive goal, is happening through mandates, not through local initiatives.

It is in the best interests of almost everyone that school systems become more responsive. It is especially in the best interest of professional groups such as the NCTM that have proposed important changes. The more responsive the system, the better the chances for new ideas, such as the NCTM's curriculum recommendations for the 1980s, to be accepted.

Sir Winston Churchill once said that "democracy is the world's worst form of government except for all of those other forms that have been tried from time to time." Another word for democracy, as applied to schools, is *responsiveness*.

The Concept of the Responsive School System

The concept of the responsive school system is based on four major assumptions:

1. Schools are infinitely improvable and efforts for school improvement should continue even during periods of declining resources. (Leaders with fixations on the status quo will not find this process useful.)

2. There should be widespread involvement in deciding what improvements should be made in schools.

3. The responsibility for planning and implementing high-priority improvements should be clearly allocated to members of the professional staff, who will be held accountable for the results.

4. Once an improvement has been agreed on and made, the results should be widely publicized.

Some elements of this system are in operation in many districts throughout the country. The total concept, however, has been implemented in only a few places. The following characteristics of this system differentiate it from other similar systems:

1. It emphasizes the setting of priorities—deciding what to do next to improve each school in a district.

2. It is responsive to input from citizens as well as staff at each stage of the process; yet it assigns professional responsibility to the professional staff.

3. It operates in a consistent manner at both the district and individual school levels.

4. It provides for action plans to be tied both to district-level and building-

level priorities. School improvements, then, are made in accordance with major needs as seen by staff and school patrons.

5. It is less complex than most processes for school improvement. The process consists of six steps that operate in a similar manner at both the district and the building levels (see fig. 6.1):

a) Asking the people—assessing needs and strengths (the responsibility of the school improvement committees)

b) Setting priorities (at the district level, the responsibility of the school board after considering the results of the study of needs and strengths and after hearing the superintendent's recommendations; at the individual school level, the responsibility of the principal with appropriate involvement of the faculty)

c) Developing action plans (the responsibility of the professional staffs at school and district levels)

d) Implementing the action plans (the responsibility of professional staffs; monitoring progress is the responsibility of each principal)

e) Evaluating (the responsibility of the professional staff, with advice on the formation of questions from pupils, parents, and other patrons)

f) Reporting (shared responsibility of the professional staff and the school improvement committees)

Description of the Responsiveness Process

The Role of the School Improvement Committees

In general, the school improvement committees are the school system's "experts" on how the staff, pupils, parents, and community leaders view their schools, and they include members from each group.

The specific role of the district-level school improvement committee should be defined annually by the board of education in a charge from the board to the committee. Each individual school's committee should also receive a charge from the principal of the school. These charges define the role of each committee so that time is not wasted by the committee in discussing what it is supposed to be doing. In general, the role of the school improvement committee is as follows:

1. Identifying as accurately as possible those improvements in the school or school system that are most widely desired by parents, other adults, students, and staff (In a responsive school system, the planners need to know what the people would like to see improved.)

2. Making recommendations to the superintendent and the board of education and, at the individual school level, to the principal regarding the district's and the school's priorities for school improvement (These recom-

mendations should be supportable by the data collected and analyzed during the study of needs and strengths.)

3. Keeping its membership and constituencies informed regarding the action plans developed by the staff, the staff's progress in implementing these plans, and the evaluative information regarding the effectiveness of the plans

The Role of the Board of Education

The role of the board of education in the responsiveness process is as follows:

1. Approving a policy statement on recommendation of the superintendent committing the district to implementing a system to increase its responsiveness

2. Establishing a policy regarding the composition, purpose, and role of the school improvement committees (In some states this has been done for the board through legislation or mandates.)

3. Considering recommendations from the superintendent and from the school district's improvement committee on priorities for school improvement for the district and defining, by resolution, which priorities are to be addressed

4. Making such budget adjustments as may be possible so that resources will be available for activities for school improvement

5. Keeping its members and constituencies informed regarding the action plans developed by the staff, the staff's progress in implementing these plans, and the evaluative information regarding each plan's effectiveness

The Role of Staff

During the "ask the people" process and before priorities have been designated, staff members serve as the school improvement committees' most knowledgeable resources regarding which potential priorities are considered most important by the faculties and students. They assume leadership roles in conducting that portion of the survey of needs and strengths that assesses the opinions of staff and students.

Once priorities have been set, district-level supervisors and school-level department chairpersons become designated leaders of task forces for those school-improvement groups dealing with curricular change and instructional improvement.

Each individual school's procedure for identifying its own unique priorities will vary depending on the way the school is organized for decision making and the administrative style of the principal. Several techniques have been developed for involving large numbers of people in the process of defining priorities. One of the most commonly used, the Delphi process, is described extensively in the professional literature. This process, when used at a meeting of parents, staff, and student leaders, will enable participants to arrive at a consensus (or near consensus) on which of the many possible priorities for improvement should be adopted. It can also be used as a means of obtaining a consensus through a series of printed surveys.

Once a school's priorities for improvement are identified, faculty members will serve on task forces to develop, implement, and evaluate the school's plans for improvement.

Staff members serving on the district-level school improvement committee will provide the committee with progress reports on how the process for planning action is proceeding at both the district and local school levels. They will also represent the staff's views on such important matters as the design of the district's evaluation system as that system relates to the priorities and the design of the "report to the people" process.

The superintendent, as the executive of the board, provides professional expertise to the school improvement committees so that they can do a professionally acceptable job in representing the opinions of the public and staff and in basing recommendations on information. The superintendent (or a designee) meets regularly with the district-level committee to provide it with information and technical assistance. The principals perform similar roles with their individual schools' committees.

A major responsibility of the superintendent is to develop a district-level school-improvement plan for each of the priorities identified by the board. A major responsibility of each principal is to develop a plan of action for individual schools for each district-wide priority and for each individual school's priority that might be different from district-wide priorities.

Once plans have been developed, the superintendent and principals assume responsibilities for their respective implementation and evaluation. Many responsibilities for implementation and evaluation will, of course, be delegated to others.

The Steps of the Process

The steps of the process are shown diagrammatically in figure 6.1.

Step 1—Assess Needs and Strengths

This step is essential if the members of the school improvement committee are to be able to represent views other than their own. The step has two components: "ask the people" and "analysis of information from program evaluation."

Ask the People. The ask-the-people process is similar at both the district and the individual school levels. Obviously, however, the target groups differ. School-level surveys are aimed primarily at the parents, staff, and

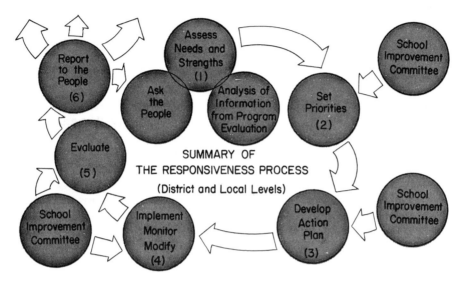

Fig. 6.1

students of an individual school. District-wide studies are aimed at the attentive public of the total district. Such studies may include information from the committees of individual schools. In fact, individual schools' committees may be asked for recommendations on district-wide priorities.

An effective ask-the-people process will have the following characteristics:

- Information regarding perceived strengths and needs will be gathered. Recommendations of priorities related to strengths (see fig. 6.2) will be for the purpose of encouraging excellence—district-wide or individual school programs that can be a source of pride for the public, the staff, the board, and the students. Recommendations of priorities related to needs (see fig. 6.3) will be designed to strengthen perceived weaknesses in the system or school. The committee believes that both kinds of information are important in building a high-quality school or system.

- The lists of items for both surveys will be determined by the school improvement committee on the basis of suggestions made by people at special meetings called for that purpose.

- The survey is supplemented by other sources of information regarding public opinion, such as minutes from public meetings at which significant issues are discussed, petitions, written statements from pressure groups, and evaluative reports from the accreditation association's visiting teams.

- The surveys will be continuous. Initial surveys are for the purpose of providing a portion of the information on which decisions for setting

---

Survey of the Washington School's Strengths—Sample Items

At an early date the Washington School's School Improvement Committee will be making recommendations regarding how the school can best be improved in the next two years. Some of the committee's recommendations will be related to the school's needs; others will be related to the school's strengths. The school improvement committee believes that efforts to improve on the school's strengths will result in the development of several "islands of excellence"—programs that can become a source of pride for parents, staff, and students.

On this form we have listed several of our school's strengths that have been suggested by students, teachers, staff, and parents.

If you disagree that some of these items are strengths, please show your disagreement by placing a "D" in the blank provided.

Of the items that you see as strengths, choose the *three* that you would most like to see developed into "islands of excellence." Place an "X" in the blank to the left of each of these three items.

_____

_____ 1. The quality of instruction in the following subject:

      _____ reading   _____ physical education

      _____ art   _____ music

      _____ mathematics   _____ homemaking

      _____ science   _____ industrial arts

      _____ social studies   _____ dramatics

_____ 2. Interscholastic athletics

_____ 3. Student council

_____ 4. Counseling

_____ 5. Overall competence of the teaching staff

_____ 6. Programs for gifted and talented students

_____ 7. Programs for handicapped students

_____ 8. The dropout prevention project

Fig. 6.2. Example of a survey of a school's strengths

priorities can be based. Once priorities have been set, subsequent surveys can be designed to determine in greater detail the kinds of program improvements that are desired.

Most school improvement committees will need professional assistance in designing, conducting, and interpreting a survey of needs and strengths. If

Survey of Needs, Centerville School District—Sample Items

At an early date your board of education will be making decisions regarding how our schools can be improved in the next two years. We hope that you will assist board members with these very important decisions by sharing your opinions regarding needed improvements in our schools.

On this form we have listed several school improvements that have been suggested by students, teachers, board members, parents, and other citizens. Please rate each suggested improvement on a scale of 1 to 4, as follows:

1. This improvement is not needed.
2. This improvement is needed but should not be acted on this coming year.
3. This improvement is needed and should be acted on during this coming year.
4. It is extremely important that this improvement take place as soon as possible.

I am a _____ parent of a pupil in the public schools; _____ parent of a pupil in nonpublic school(s); _____ nonparent adult; _____ student; _____ administrator; _____ teacher or specialist; ___ support-staff member; _____ other.

_____ 1. Improve instruction in basic skills, especially in the following (check one or more):
    _____ Reading
    _____ Writing (including spelling, punctuation, grammar)
    _____ Computing
    _____ Critical thinking, decision making, and practical problem solving
    _____ Other (specify)

_____ 2. Expand intramural and athletic programs to encourage more participation by girls.

_____ 3. Improve career education programs.

_____ 4. Improve and expand vocational education programs.

_____ 5. Improve learning opportunities for gifted and talented pupils, especially in the following (check one or more):
    _____ Art
    _____ Music
    _____ Dramatics
    _____ Science
    _____ Math
    _____ Language Arts (literature, creative writing)
    _____ Social Science
    _____ Physical Education and Athletics
    _____ Other (specify)

_____ 6. Expand and improve school buildings through a bond issue.

_____ 7. Improve students' morale and discipline

_____ 8. Other (specify)

Figure 6.3 Example of a survey of a school district's needs

such assistance is not available from district staff members, it may be obtained from a university service bureau or sometimes from the state or regional department of education.

Analysis of information from program evaluation. The most commonly available type of information from program evaluation is information from the district's or individual school's testing programs. Other important information may come from the district's evaluation of the extent to which there is congruence among what is being taught, what is being tested, and what the formal curriculum guide says is being taught. (See Fenwick English's *Quality Control in Curriculum Development,* available from the American Association of School Administrators, 1801 N. Moore St., Arlington, Va.)

As an important step of the assessment of needs and strengths, existing information from program evaluation should be analyzed. Since this analysis is a professional task, it should be done by a professional person who has training and expertise in program evaluation. Summaries and interpretations of existing information should be presented to the school improvement committee.

On receiving information from program evaluation, the committee compares what the people think about the need for improving various programs with existing information regarding the effectiveness of these programs. Sometimes the people's opinions will be supported by the evaluative information. The people may hold the opinion, for example, that the district's mathematics program needs to be strengthened, and this opinion may be reinforced by disappointing test results. Then the rationale for designating the improvement of mathematics instruction as a priority would be strengthened. In other situations, there may be a discrepancy between what people think and what information from program evaluation supports. The people may feel, for example, that the district's reading program is weak, but evaluative information may indicate that it is strong. In such situations the real need may be to improve the dissemination of information regarding pupils' performance rather than modify or replace the program.

A completed study of needs and strengths, then, consists of two parts—asking the people and analyzing information from program evaluation. On the basis of this information, recommendations for setting priorities can be made.

Step 2—Set Priorities

At this point the district-level school improvement committee analyzes the results of the study of needs and strengths and reports its findings and recommendations to the board of education (see fig. 6.4). The most important task of the district's committee is to recommend to the board which priorities for school improvement should be adopted. These recommendations should be justified by the data from the study of needs and strengths. It is the job of the board of education to make the final determination of the district's priorities.

Following a discussion of the report, the board refers the report to the superintendent for further study and recommendations. At a subsequent meeting the superintendent responds to the findings and recommends that the district adopt from two to six of the priorities identified as important. The board then acts on the superintendent's recommendation, officially adopts the district's priorities, and charges the superintendent with the task of developing action plans for implementing each priority they have authorized.

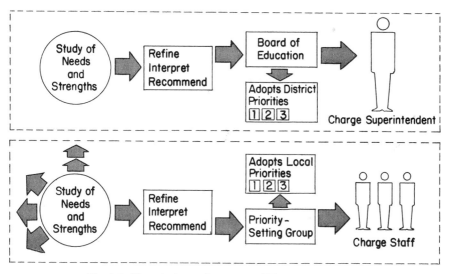

Fig. 6.4. The priority-setting process (What do we do first?)

Meanwhile, a similar priority-setting process is taking place at the local level. At the district level the leadership for priority setting has come from the superintendent and district-level staff; at the local level this leadership comes primarily from the principal. The principal may, however, delegate much of this leadership to teacher-leaders or supervisors.

In some schools the priority-setting decision will be made by the principal on the basis of information that has been gathered and interpreted by the school improvement committee. In other schools the decision may be made by the committee, the faculty, or a faculty-student-parent group formed for this purpose.

Regardless of who makes this decision for setting priorities, the principal and the committee should insist that the decision be justifiable from the information that the committee has gathered and interpreted. A committee will not be easily motivated to do further ask-the-people surveys if the results from the first one are ignored at decision-making time.

A priority-setting decision is most easily made when—

1. the survey shows that all major groups questioned see the potential priority as important; and
2. the school's plan of program evaluation shows that the program being considered for priority is yielding disappointing results; or
3. all major groups questioned on the survey of strengths designate the same program a potentially excellent one.

At least one priority for improvement should be chosen from the results of the survey of strengths. Thus, one group of people in a school will always be working to improve the extent to which good programs become excellent. For example, a school might define one of its priorities for improvement as expanding the extent to which its diagnostic and prescriptive teaching—a successful practice in its special education, science, and mathematics courses—becomes common practice in other areas. Another school might choose to strengthen an already strong program, such as its gifted and talented program or its experiential education program, so that that program will become an "island of excellence."

Once the school's staff adopts its unique priorities, the principal will submit the school's list of priorities to the superintendent for final approval or modification. On approving the school's priority list, the superintendent charges the principal and staff with the task of developing action plans that will define the specific program improvements the principal and the school staff will make.

By this time, each school in the system will have defined from three to six priorities for school improvement. Some of these priorities will have been designated by the board of education on the basis of information provided by the district-level school improvement committee. Other priorities will have been set by the school's faculty on the basis of information provided by the individual school's improvement committee. One school, for example, found itself, as a result of the priority-setting process, with the following district-wide priorities (applied to all schools):

1. Improve the school's reading program.
2. Strengthen and expand programs for gifted and talented pupils.
3. Improve students' morale and discipline.

It also had the following priorities unique to itself:

1. Improve cafeteria service.
2. Expand and improve athletics and intramurals (already successful programs).

3. Strengthen instruction in mathematics for the school's least able mathematics students.

Step 3–Develop Action Plans

At this point, the principal, with the advice and support of the school improvement committee, forms one task force for the planning and implementation of each priority for school improvement.

A task force differs from a committee in several important ways. Committees typically are formed for the purpose of studying a topic and making recommendations. Task forces are formed to plan and implement improvements. The task force is the key organizational unit in the process of improving schools. It is the task force that does most of the work. Without the task force, plans are not implemented; nothing much happens.

The membership of a task force depends, of course, on the nature of the task. A task force organized to improve students' morale and discipline might, for example, be composed of a counselor, the sponsor of the student council, student leaders, the assistant principal for pupil personnel, and two or three teachers who are highly respected for the extent to which they are able to promote high morale and productivity in their programs. A task force to strengthen instruction in mathematics for the school's least able mathematics students would probably be composed mostly of mathematics teachers. Parents and nonparent community leaders may be invited to serve on task forces if they have talents of value to the group or if they represent groups that wish to be kept informed of the progress of the task force.

The first duty of each task force is to develop its action plan. Each plan should contain at least four essential elements:

1. Objectives for program improvement
2. A list of activities related to each objective
3. An indication of who on the task force is responsible for the accomplishment of each activity
4. Target dates for the completion of each activity

A fifth element, how each objective is to be evaluated, is also essential. The task force may, however, take longer to define its plans for evaluation than to define its first list of activities.

Figure 6.5 shows a sample work plan developed by a task force working on implementing a teacher-advisement program. One planning sheet should be developed for each objective related to the priority.

Through the task-force system, anyone in the school or community who wishes to participate in making the school a better place will have an opportunity to do so. Schools that have organized such task forces report improved morale of faculty and students as the people in the school realize

Task Force on Teacher Advisement

Objective: Improve communications between teacher and pupils
                  and provide each pupil with faculty support in personal
                  problem solving and decision making

Responsibility: Jim Hanson, assisted by: Marie Davis, Jo Ann Brown,
                                            Bill Buckner, Harold Jensen

Activities	Target Date	Responsibility
1. Provide members of task force with information regarding successful programs	Ongoing	Jim
2. Organize visits to other schools	Sept. 20 Oct. 15 Nov. 15	Bill
3. Design pilot project for Jefferson High School	Dec. 1	Harold and Total Task Force
4. Design evaluation plan for pilot project	Dec. 1	Jo Ann
5. Plan and conduct in-service programs for advisors of pilot project	Oct. 15 Nov. 15 Dec. 5 and Ongoing	Marie
6. Keep faculty, parents, and students informed of projects	Ongoing	Jim and Marie

Fig. 6.5. Sample of the work plan of a task force

that they have power and influence. They are actively involved in shaping the institution that in turn will shape their behavior and that of their pupils.

The principal assists the task forces in the development of the action plans. Once completed, the plan is approved by both the task force and the principal. The principal then gathers all the school's action plans together, adds some introductory and background material, and submits the resulting document to the superintendent's office as that school's overall plan for school improvement.

A member of the superintendent's staff summarizes the contents of the plans from all the schools, adds the district office's plans for staff development, evaluation, and support for program improvement, and then publishes the resulting document as the district's action plan for school improvement.

The district-level action plan contains one section for each district-level priority that summarizes what each school plans to do about that priority. Another section of the district-level plan describes what the superintendent's office will do to support the improvement efforts of the individual schools. A final section contains recommendations to the board of education on policy or budgeting.

This plan is reported to the board and the people as a part of the superintendent's budget recommendations. In this way the superintendent presents the recommended budget to the people as a response to the people's recommendations of priorities for school improvement. The superintendent is saying to them something like this:

> We have asked you for your recommendations for improving our schools, and you have responded by telling us that you want them improved in three important ways. Your board charged the professional staff with developing plans for implementing the changes you recommended. We have completed this planning task, and we are now ready to implement the authorized improvements. The provisions for implementing these improvements are included in the district's budget.

At this point the board and the community leaders will begin to realize what it means to operate a school district that is responsive to the people.

Step 4—Implement, Monitor, and Modify

The staff of each individual school now begins to implement its action plans. Each plan calls for the allocation of responsibility for achieving specified tasks. The monitoring of progress is the responsibility of each principal. As the implementation of each plan proceeds, it will probably become necessary to modify the plan by adding activities, altering time lines, or changing allocations of responsibility. The principal makes these modifications with the knowledge and approval of the superintendent or the superintendent's representative.

During this period it is especially important that the evaluative activities be accomplished.

Step 5—Evaluate and Get Feedback

The evaluation of the individual school's and the district's action plan is of two types, formative and summative.

Formative evaluation is ongoing. An action plan may be modified on a day-to-day basis as information regarding its effectiveness becomes known. Some of this evaluation is very informal, almost intuitive. All of it need not be recorded. Formative evaluation is an analysis of *what is happening.*

Summative evaluation is formal and usually statistical. It is an analysis of *what has happened* in reference to the questions that were asked at the time the action plan was developed.

Each principal and task-force leader reports the available evaluative information regarding the various action plans to the school's staff, students, and parents. At this point the principal and leaders are saying something like this to the students, staff, and parents:

> We have been working for a year on the school improvements that you thought were important. As a result of this work, the following good things have happened:
>
> _____
>
> _____
>
> We have, however, many unfinished tasks. These will be built into next year's action plan.

Each principal also provides the superintendent's office with reports on the effectiveness of the school's action plans. This information from all the district's schools is then combined by the superintendent's staff into a comprehensive evaluative report to the district's school improvement committee, the board, and the people. The superintendent, when presenting the report, can say something like this:

> You have told us how you wanted your schools improved, and we have developed plans for improving the schools in all priority areas. You reviewed these plans with us earlier this year. Now we are ready to tell you what happened. . . .

At this point the board and interested community leaders can readily see that their school organization has been responsive to their wishes. The support of the public and the board for further improvements in the next budget cycle will be strengthened.

As a result of the evaluative information, the superintendent may recommend that all or some of the district-wide priorities be continued for another year or two-year cycle.

While the staff has been implementing its action plans, the school improvement committees will have been further assessing the opinions of the community, staff, and students regarding needed improvements. Their study may provide evidence that the board should consider adding one or more priorities to its list. Thus, through the priority-setting process, the people speak through their school's improvement committee to express their opinions regarding important school issues.

It is at this point that the feedback responsibility of both the district and the local school improvement committees becomes critical. It is important that evaluative information, both positive and negative, be reported to the responsive publics through the same channels that were used to collect information on needs and strengths. Otherwise, the responsive publics, which include staff members and pupils, may never learn that their views had been heard and acted on.

Here, too, the individual school's committee might make some specific recommendations to the district-level committee regarding the district's priorities to be recommended for the next cycle. If several schools are working on the same or similar priorities, the board of education may wish to designate those priorities as district-wide priorities.

Conclusion

The system described in this chapter is consistent with the various theories and strategies recommended by experts on the change process quoted in previous chapters.

Because of limitations of space, some components of the system are not described in complete detail; there are gaps in the system as presented. However, a blueprint is presented for building a responsive school system, a blueprint that, if followed, will result in improved programs and enhanced community support.

This system is based on the major assumption that the attentive public—the people who work and learn in schools and others actively concerned with school affairs—should be actively involved in the process of change. Key organizational units for encouraging involvement are the local and district-level school improvement committees and the task forces. Six major steps in the responsiveness process have been defined.

This system will, of course, operate very differently in each community. In schools with strong traditions of involvement of community, staff, and students in school improvement, the system will work smoothly. The responsibility will be widely delegated and the work load broadly shared. In school districts with traditions of authoritarianism, the system will work less well.

Such a system would well serve the interests of mathematics teachers because it would—

1. provide such a group with a systematic process whereby the improvement of mathematics programs could become the priority of a district or an individual school;

2. provide such a group with ready-made channels of communication with the superintendent's office, the board, and attentive publics;

3. provide procedures for such a group to obtain financial and personal support for its efforts at improving and evaluating its programs.

Chapters 7 through 13 give specific suggestions for mathematics teachers who are seeking effective ways to implement the NCTM's recommendations for a curriculum for the 1980s. If followed, these suggestions will enable a task force on mathematics improvement to work effectively within the overall structure recommended in this chapter.

PART

Changing Mathematics Programs

Education which is not modern shares the fate of all organic things which are kept too long.

ALFRED NORTH WHITEHEAD

The ideas and concepts on change presented in Part One provide the basis for the chapters in Part Two. The following chapters consider what is known about change, changing teachers' behaviors, and changing schools and then apply this knowledge to changing mathematics teachers' behaviors and changing mathematics programs. In particular, the motivation for change is to implement the NCTM's *Agenda for Action: Recommendations for School Mathematics of the 1980s*. Curricular reform is certainly not new to the mathematics community. In the 1960s large-scale efforts for curricular reform were supported by the National Science Foundation—for example, the extensive work of the School Mathematics Study Group (SMSG). SMSG's efforts and those of others were hailed as the "new mathematics" with great fervor and zeal by mathematicians and educators alike. In recent years, the effects of this new mathematics have been somewhat discounted. Nevertheless, some long-lasting effects of these unsophisticated efforts at curricular change are evident. Contemporary textbooks have virtually "institutionalized" some of the content and methodology suggested as new mathematics. In addition, mathematics education—the teaching and learning of mathematics—has come into a field of its own. These phenomena can be considered a residue of those earlier efforts at massive curricular change.

It is now some twenty years after the introduction of the new mathematics, and mathematics educators, mathematicians, and educators have perceived the need for a renewed effort at curricular change to implement the curriculum of the 1980s. However, the professional community has grown in sophistication with respect to the change process. We know it is necessary to identify what is to be changed, but knowing the "what" is not sufficient. We must also attend to "how" to change. Certain strategies for change are more effective than others, and this time in our efforts at curricular change we can knowingly consider and include these effective strategies.

The earlier chapters provided us with guidance with respect to change in general. In Part Two we shall consider change with particular application to the teaching and learning of mathematics. Coburn's chapter provides specific structures and strategies for assessing needs, attitudes, and settings for changing a mathematics curriculum. Rowan and Morgan build on this theme by describing and illustrating the steps involved in designing a plan for change. Taylor adds to the specifics of assessment and plan design with effective techniques for developing the support of other significant persons in implementing the desired change. Finally, Retson identifies ways to monitor and evaluate change. Each of these chapters is written with specific guidelines on how to change—that is, to reform and improve—the mathematics curriculum.

Change is not—and should not be—a lonely business. The teacher in the classroom, the mathematics department chairperson, the principal of the school, the mathematics supervisor of the district, and the mathematics educator at the university will each find it difficult if not impossible to implement change alone. The support, approval, cooperation, advocacy, and involvement of others is critical. Identifying and locating the different kinds of power and support bases are critical first steps in assessing and diagnosing readiness and receptivity for change. Using, matching, and adapting to local parameters known models for change can facilitate our efforts. Knowing how and when to monitor or intervene can also contribute to the likelihood of success.

These chapters have been written to mathematics educators—both individually and collectively—in order to increase their knowledge and awareness of the processes of change. Specific strategies of change, both in general and with particular application to mathematics curriculum and also both theoretical and practical, have been presented. As we go about the business of changing curriculum this time, we not only must know our curriculum but also must know how to change it effectively.

7

Assessing the Status Quo

TERRENCE G. COBURN

ASSESSING the current status of mathematics instruction in a school is the first step in a process of change. The major purpose of the change is to improve learning. Recent developments in education have increased the potential for testing to help improve instruction. Objective-referenced tests and large-scale assessments have helped to show that data from students' performance on specific instructional objectives are useful in determining instructional needs and strengths.

The term *assessment* refers to any procedures or processes employed in collecting information. *Evaluation* involves relating descriptions of behavior to other variables or making comparisons among various descriptions of behavior. A test is merely one sort of measurement device. An assessment should be more than just the use of a test.

An assessment can be a major contributor in a process of planned improvement. It can also have some negative aspects if employed improperly. For example, giving too much emphasis to the attainment of minimal objectives in an assessment can have a debilitating influence on the mathematics curriculum. An assessment can also provoke controversy when it is perceived as a political tool concerning the adequacy of the staff or the allocation of funds. A successful assessment requires effective leadership to steer a course through the stormy issues of testing and accountability. Prudent judgment, good communication, and careful planning will help reduce the negative aspects and increase the positive features of an assessment.

Aside from the controversy it often engenders, an assessment seems to be a rather straightforward and routine process. One begins by specifying what is to be looked at and then obtaining information-gathering instruments. The data are collected and analyzed, the results are interpreted, and decisions are made. Why is it so difficult to conduct a proper assessment? This chapter attempts to provide some grist for thinking about assessing a school's mathematics program. It is not a detailed, step-by-step, "how to do it" chapter. Circumstances and resources vary too widely to warrant such a

85

cookbook approach. Rather, the chapter underscores considerations raised by the author from experience with district-wide and other large-scale assessment efforts.

The focus of this chapter is a hypothetical school about to assess the status quo of its mathematics program. "The school" will be used as a placeholder for whoever is conducting the assessment; it may be a large district or a single school. The reader can apply the generalizations to the specific parameters of the situation. Of course, the size of the school, the scope of the assessment effort, and the available resources will dictate practical limitations, but certain considerations should be applied regardless of the size of the assessment effort. It is also assumed that the school in question has recently done very little to collect data or evaluate its mathematics program. This is an initial effort for the school as it begins to take a close look at its program of mathematics instruction.

The Assessment Committee

A school's administration can begin the effort at assessment by forming an assessment committee. In this chapter, "the assessment committee" or "the committee" will refer to whichever committee would be relevant in the particular local arrangement. The reader should concentrate on the nature of the tasks to be accomplished and not the mechanics for carrying out those tasks. In certain situations the assessment committee could conceivably consist of one person. The reader should realize, however, that an assessment of any significance is a major undertaking; if the staff is limited, provision should be made for whatever assistance can be arranged. The assessment of the status quo is of fundamental importance to the mathematics program of the school, and this importance cannot be overstated. Think of the assessment as the foundation to a future alteration of the instructional program. If the foundation is not laid properly, the subsequent program may be weakened. The leadership and the representativeness of the workers in this effort are crucial. There should be representation, whenever possible, by classroom teachers from grades K–12, administrators, mathematics educators, test specialists, and guidance counselors. When a major function is not represented by a person on the school's staff, then input for that function might be arranged through the services of an outside consultant.

Certainly the membership of an assessment committee can change as the nature of the task at hand changes. An elaborate committee structure is not necessary for an assessment, but the committee model is used here as a convenient vehicle for presenting thoughts on conducting an assessment.

Clearly, the committee must be given adequate time to complete its task, and its effort must be well funded. A firm commitment from the school's administration and a clear and concise statement of purpose are necessary.

The committee's sense of what it is to do will shape its decisions and recommendations. Less effort will be wasted and fewer conflicts will arise with the intentions of the school's leadership if the committee's charge is thoroughly understood by all concerned at the beginning.

Special emphasis should be placed on communication. The committee's major role is the planning and carrying out of the assessment, but it should also keep other significant persons involved or aware of developments. As noted earlier, an assessment can be a controversial project. Confusion will exist about minimal-competency testing, accountability, reporting individual students' achievement, and so on. School colleagues and citizens in the community must not feel that they are being kept in the dark about the committee's intentions and progress.

Finding the Focus of the Assessment

An assessment of the status quo can begin with a preliminary examination of current issues and forces. The assessment needs to find a focus—an area or issue about which the information will be most useful for the decisions that will be made about any possible changes. Preliminary data gathering and discussion are necessary to determine if a change is needed. If a change seems desirable, then this discussion can speculate about the type of needs that are anticipated and which needs might be of higher priority. The resources needed to conduct the eventual assessment can be examined in relationship to the anticipated importance of the data. Four major elements in this preliminary examination of the status quo are these:

1. An examination of the issues facing mathematics education today and a preliminary discussion of the relevance of these issues to the current mathematics program
2. An analysis of the policies of the school and the climate of both the school and the community regarding possible changes in the program
3. A review of the goals of a high-quality mathematics program for the future and a preliminary indication of the existence of these goals in the current mathematics program
4. An analysis of the resources available to conduct the anticipated assessment and a determination either to seek additional resources or to cut the scope of the assessment

The assessment committee must examine the issues and emerging trends confronting our school mathematics programs today. Some concerns for mathematics education are listed below. If one or more of these issues is deemed crucial by the assessment committee or other school leaders, then the assessment can be shaped to obtain relevant data in the areas related to those particular issues.

1. Problem solving is emerging as one of the major goals of mathematics education.
2. The existence of the electronic hand-held calculator enables more people to make better use of mathematical applications.
3. The computer pervades much of our daily living and should be understood by more of our citizens.
4. Statistics and probability are emerging as an important body of content containing skills and understandings needed by an increasing number of persons, both professionally and in everyday life.
5. The metric system, now the world's principal system of measurement, should be the system of measurement taught, but instruction in the customary system is still needed.
6. A recent movement calls for de-emphasizing fractions and shifting the traditional sequence of fractions and decimals to give more emphasis to decimals. The committee should give careful attention to this trend and be prepared to give full explanation for any objectives that represent a significant departure from the present program of instruction.
7. Concern arises today for the apparent mathematics avoidance among girls in secondary schools. Female students may be limiting their future educational and career choices by avoiding mathematics courses.
8. "Return to the basics" is the slogan of a major educational trend today. The mathematics education community is concerned that this trend will place too much emphasis on rote computation and lead to debilitation in other areas of priority of the mathematics program.
9. Minimal skills and minimal-competency testing are receiving considerable national attention.
10. Educating the gifted and talented student is causing many teachers to look at so-called enrichment topics and acceleration of courses in the mathematics program.

Issues identified as high-priority areas of concern should be examined in relation to any forces within the school and the community that would impinge on any decision to bring about changes in those areas. An analysis of the forces that appear to be promoting change and the forces that may restrain or militate against change would be helpful in shaping the eventual assessment. This is not to say that the assessment should be designed to favor one position or another but rather that the data collected will be helpful in pointing the direction toward desirable change.

The assessment committee should gather information and testimony about the policies of the school and the structure of the institution. Policies regarding minimal competency, programs for the gifted, and so on could have an effect on possible change, and these data belong in the assessment of the status quo. Unintentional or unwritten policies may exist, such as a lack of encouragement for girls enrolled in higher mathematics classes. An impending move from a junior high school situation to a middle school arrangement could focus the assessment on the possible needs of students in this new arrangement.

The climate of the community should also be considered. Concerns over test scores, the use of technology in teaching, perceived attainment of basic skills, and other areas of interest need to be examined. The career goals of the students could be surveyed with assistance from the home. Enrollment patterns as compared with those in private schools could be examined. A follow-up on college students and their mathematics requirements and successes (or failures) could be undertaken.

The climate of the school, including teachers' attitudes, skills, and knowledge should also receive a preliminary examination. Some indication should be obtained about whether potential change will receive the support of the school's staff (see chapter 9, "Developing Support for Change"). Preliminary data may have to be gathered by school leaders for use in obtaining support for collecting further data in a controversial area. Suffice it to say that if the staff is hostile to a possible change, they are unlikely to view assessment data in an unbiased manner.

A survey should be made of all possible sources of relevant data currently available in existing school records. Attendance and patterns of course enrollment can be reviewed. Standardized test scores as well as other indicators of students' achievement, such as departmental exams, may be available. It may not be necessary actually to gather and review these data at this time, but it is worthwhile to make a note of what is available.

Another factor of the climate of the school is the current instructional practices in the mathematics program. Teachers could be surveyed in such areas as (1) their attitude toward teaching mathematics, (2) the amount of time used daily to teach mathematics (in the elementary school), (3) the amount of time spent teaching specific topics or skills (in both elementary and secondary schools), and (4) the types of instructional materials used. Information on instructional practices, especially the time devoted to instruction on specific skills, will be useful in interpreting the results on students' performance. This aspect of the assessment will be discussed further following a discussion of the goals and objectives of the mathematics program.

It is clear that assessing the status quo is more than gathering data on students' achievement and attitudes. An assessment is the gathering of all pertinent information. The questions that the assessment is designed eventually to answer may require input from the school's records, testimony from school leaders, surveys of teaching practices, and input from the community.

Few schools have the resources to conduct a wide-range assessment of both process and achievement over all grade levels at one time. The school leaders must select some high-priority area of concern in which needs are most likely to show up in the assessment and in which there is a high probability that change is possible, feasible, and operational. Among the several practical limitations that need to be considered are the following:

- *Time.* There is not enough time to measure everything of interest. Even if a substantial amount of time were available, the students should not be subjected to more testing than is absolutely necessary.

- *Relevence.* This is not the time to pose too many "interesting" research questions. The results must be relevant to the improvement of instruction. The committee might well challenge each test item on the basis of whether or not the results can be interpreted in terms that will help teachers understand what to teach or how to teach better.

- *Feasibility of change.* Look ahead to the usefulness of the results; if the possible changes indicated by the results are not feasible, then there is some question about whether resources should be expended in collecting the data.

- *Sophistication of the data-collection procedures and devices.* Some hard decisions must be made in this area about what is and is not measureable depending on the degree of cleverness or state-of-the-art capability of the procedures to measure certain highly desirable but complex behaviors.

- *Money.* It costs money to collect and interpret data. This would be a good time to count the costs and make sure the budget is adequate. The budget may need to include funds for training a cadre of data collectors.

Goals and Objectives of the Mathematics Program

After the preliminary examination of forces and issues has given an initial direction to the assessment, the committee next turns to the instructional objectives that will be measured by the assessment.

The committee will, in all likelihood, not be starting with a blank slate. They will wish to retain some goals of the current program and begin to assemble new goals and objectives from other sources. They should not be restricted to the school's current program. For example, they may decide that some goals in the area of statistics need to be adopted. If a statistics class is not currently taught in the school, the assessment will probably demonstrate a need for such a class.

The goals and objectives assembled by the committee will help shape the specifications for the future data-gathering effort. These should be statements of desirable outcomes forged from debate over the important question, "What mathematics is of most importance to learn?" Now is not the time to consider whether or not an objective is measureable; that can be decided later. The task now is to help document what ought to be the ingredients of a first-rate mathematics program. This is not to say that the committee shouldn't be practical and attempt to be limited by what is feasible. There is no sense in wasting time on an effort that cannot possibly bear fruit. The committee should govern their current activity by the notion that if the objective is teachable and learnable, then it can be measured in some way.

One might expect that the description of a high-quality mathematics program would have been formulated and adopted a long time ago. There is, however, no established or widely accepted set of goals for mathematics education. There are too many points of view regarding curriculum, psychology, and other varied interests to allow for unanimity. Thus the goals of mathematics education will reflect local interests, varying from place to place, and will also change over time. A study of the history of mathematics education reveals that its goals change with the ebb and flow of different forces and issues that impinge on our society in general and our schools in particular (National Council of Teachers of Mathematics 1970). World wars and economic depression are two obvious forces that have affected our goals in the past. Technological growth, inflation, and declining test scores are shaping our goals today.

Goals represent the biggest targets of mathematics instruction. They indicate a school's overall view of the question "Why study mathematics?" The goals reveal a mixture of purposes—practical utility, cultural interests, preparation for subsequent study in the sciences, and so on. The goals reflect the influence of the professional mathematician and the mathematics educator. It is also apparent that the goals must reflect the plans and aspirations of the community for their children. The college bound and the non–college bound must both receive the very best mathematics preparation possible.

The mathematics program should be assessed in terms of the effects of the program on the students. The objectives used in the assessment are descriptions of the behavior expected of the learner at the completion of instruction. Behavioral objectives are a rather recent development in education. Their impact has been extensive and fruitful. The committee should be made aware of such past mistakes as voluminous listings of very narrow and fragmented statements of instructional outcomes. Guidelines are not readily available for help in determining optimal numbers or preferred wording of objectives. Suffice it to say that an instructional objective should clearly and concisely describe a specific outcome of learning in terms of what the student will be able to do given certain agreed-on conditions.

Instructional objectives comprise not only a range of content categories but also different levels of cognitive behavior. Benjamin Bloom and others have been helpful in describing levels of thinking from basic recall of facts and routine processes to higher-level applications and analyses. One criticism of many lists of mathematics objectives is that they consist primarily of lower-level cognitive objectives. The committee should seek to arrive at a comprehensive set of objectives representative of the different cognitive levels. The National Asssessment of Educational Progress (NAEP) has an excellent publication describing a useful matrix of content by cognitive behavior (NAEP 1978).

Affective behavior should also be considered. The attitudes and feelings of students toward mathematics can play a fundamental role in an assessment. It seems clear that positive attitudes correlate well with mathematics achievement. However, specific guidance about appropriate affective objectives has been lacking in the mathematics education community. We have little in the way of helpful advice for the committee other than to suggest that they create a special category for affective objectives for which some assessment data will be collected. This area is examined further in the discussion of data gathering.

The committee must work hard at broadening the scope of the assessment. Textbooks and stan'dardized tests have a tremendous influence on the mathematics curriculum. The committee should carefully consider this influence and be prepared to defend goals that are not adequately supported by current texts or measured by current tests. This is not to say that new goals and new objectives are out of order; however, one should anticipate that assessment results may be lower in areas involving new goals. The community, the school's staff, and the administration will expect an explanation. Elementary school textbooks are notoriously weak in their coverage of problem-solving skills. A committee that sets a high priority on problem solving should be prepared to find relatively low attainment of problem-solving objectives by students. Standardized tests have numerous items on fractions at specific grade levels. A school that de-emphasizes its goals concerning fractions may have to defend low test scores.

Today's climate of minimal-competency testing, emphasis on basic skills, and accountability pressure requires special mention here. It is not appropriate to belabor these issues, since so much has already been written about them. However, it is worth noting that the committee will need to give careful attention to any tendency to take the road of easy, computation-oriented goals and objectives. A mathematics program cannot become excellent if it does not strive for excellence in its beginning. Pressure to look good on the test results may result in a tendency to make the objectives too easy. Watered-down objectives will not serve any desirable ends. The first step in an attempt to bring about change should be a comprehensive set of objectives across all high-priority content areas and at the different cognitive levels associated with a high-quality mathematics program.

The process of setting goals is clearly of crucial importance. The committee can begin its task of determining goals and objectives by assembling all or some of the following local materials when available:

- Current local curriculum guides or lists of objectives for the school's present mathematics program
- Copies of textbooks and other major relevant instructional materials used in the school
- Copies of all relevant mathematics tests used on a school-wide basis
- Statewide mathematics objectives or mathematics curriculum guides

The following materials should also be considered by the committee as it wrestles with the question of goals for the 1980s:

- The NCTM report, *An Agenda for Action: Recommendations for School Mathematics of the 1980s*. This document recommends that problem solving be the focus of school mathematics in the 1980s and that basic skills in mathematics be defined to encompass more than computational facility. It also makes recommendations concerning calculators and computers (NCTM 1980).
- The position paper on basic skills of the National Council of Supervisors of Mathematics (NCSM). A widely circulated and well-received document that outlines ten basic-skill areas in mathematics education. This report argues convincingly that basic skills must include more than computation (NCSM 1977).
- The NACOME report of the Conference Board of the Mathematical Sciences (CBMS), *Overview and Analysis of School Mathematics, Grades K–12*. An examination of current reform in the mathematics curriculum with specific and pertinent recommendations for the goals of the 1980s (CBMS 1975).
- *Priorities in School Mathematics: Executive Summary of the PRISM Project*. An interpretive summary from the NCTM's Priorities in School Mathematics project. The PRISM study gives guidelines and suggestions for changes in the mathematics curriculum during the 1980s. The study assessed the preferences of several referent groups for different content topics and instructional goals. The study also assessed priorities for change in the mathematics curriculum (NCTM 1981a).
- Various reports and interpretive articles of the National Assessment of Educational Progress (NAEP) concerning the mathematics assessment. For the reports, write to NAEP, 700 Lincoln Tower, 1860 Lincoln St., Denver, CO 80203. Interpretive articles have been published in past issues of the *Arithmetic Teacher* and the *Mathematics Teacher* (Carpenter et al. 1980; Coburn 1979; Fey 1979). The second

mathematics assessment was completed in 1977–78. Data are available on specific mathematical attainments of representative samples of 9-, 13-, and 17-year-olds (NCTM 1981b).

- *Critical Variables in Mathematics Education*, by E. G. Begle (1979). Findings from a survey of the empirical literature.

The goals and objectives ultimately developed by the committee should be validated by the classroom teachers. They or their chosen representatives should stipulate that these objectives are indeed descriptions of the instructional outcomes for which they are striving. When new goals are considered, teachers should agree on the worthiness of these goals. The priorities that the classroom teachers place on certain objectives are pertinent data for the assessment. It is not recommended that every objective rated as low priority by the teachers should be dropped from the assessment. The committee should be prepared to give a convincing argument about why certain "quality" objectives should be a part of the program even when the teachers do not expect most of the students to attain them. One of the major steps in this effort at assessment is a comparison of measured outcomes with expected outcomes. It is wise to secure a clear understanding of the intentions of the mathematics program prior to securing the test results.

As the committee narrows the focus of the assessment, one major caution should be carefully considered—how to protect those areas of the curriculum that are not tested. One potential effect of testing an objective is to raise its importance and to hold it in higher esteem than a nontested objective. Let us assume that some areas of the mathematics program will not be tested. Perhaps one nontested area is creativity in geometry at the upper elementary school or middle school level. If this area of the program is considered valuable, it should be protected from being eroded by attempts to bolster areas of the program that show weakness. The committee might attempt to retain acceptance of these high-priority, but nontested, areas through discussions by the staff of the ingredients of a high-quality mathematics program, tested and nontested areas alike.

Many persons would advise the committee, "Don't reinvent the wheel." Although it is clear that no one expects the committee to begin from scratch in its efforts, it should also be clear that the committee will not be able to adopt someone else's set of objectives "as is." Going through the process of modifying and specifying instructional objectives has considerable merit. The in-service benefit for the participants is enormous.

Questions for the Assessment to Answer

The major questions of importance in future curricular and instructional change should be formulated by the committee before the data-gathering procedures are finalized. The committee should pose those questions for which answers from the data will be relevant. These questions will have been expressed informally several times during the goal-setting process; now is the time to write them out. The nature of these questions will dictate the shape of the data-gathering effort. Some questions require more of a diagnostic approach than others—for example, "How do our primary aged children think when they work a subtraction fact like $11 - 7 = \square$?" A list of major questions will be helpful in deciding what type of tests to use, who to test, and when to do the testing. As the committee speculates on how their assessment questions will be answered, they can simulate the data-gathering procedures and make decisions on how to record the data and what type of technology to use in storing, analyzing, and reporting the data. Here are some sample questions:

- How well do our sixth graders perform the division algorithm at the end of sixth grade? How does this compare with tenth graders' performance on the same items?
- How well can our elementary school children measure with a centimeter ruler and a meterstick?
- What proportion of our algebra students take advanced courses in mathematics? Is this proportion the same for boys as for girls?
- How many of our middle school youngsters are ready to take an algebra course in the eighth grade?
- How much probability do our high school students understand?
- Do our middle school students like mathematics more or less when leaving ninth grade than when they entered sixth grade?
- When can the majority of our elementary school pupils successfully perform the subtraction algorithm?
- How successful are our sixth and seventh graders at solving multiple-step problems?

Gathering Data on Students

The committee begins to earn the praise and congratulations it will receive when it begins to sift through the goals and objectives and decide which behaviors to measure and what data to collect. The committee should now review its charge and thoroughly discuss the major purpose of the assessment. If the purpose is to collect information that will help improve mathematics instruction, then not all objectives need to be tested and not every student needs to be tested. Furthermore, assessing need not mean testing in the usual paper-and-pencil mode. Different assessment techniques and measuring instruments must be examined and sorted out in terms of their potential for gathering the most relevant data. Timed tests may be needed to measure students' recall of such basic arithmetic facts as $2 + 3 = 5$; attitude measures may be needed; interviews of individual students and diagnostic questions may be used to get better information on children's thinking. Students may be asked to use actual measuring tools or manipulative objects. These and other possible testing techniques should be examined in light of the questions that will be answered by the assessment. These questions are formed during the selection of the instructional objectives and goals that are most crucial for possible changes in the program.

Problem solving is an example of a hard-to-test area. If the committee wishes to establish problem solving as a high-priority goal and does not wish to be limited by textbook examples of problem solving, then it must describe those problem-solving skills that it considers important. Perhaps the committee will have to develop some measures of problem solving, no matter how imprecise, just to protect this important goal from being squeezed out by computational goals that are simpler to measure. If it is teachable, then it should be measureable to some degree. The limitations of measurement should not alter the goals of the program.

An assessment is more than testing, but one of the major data-gathering instruments used is the paper-and-pencil test. One of the committee's paramount decisions is whether to use commercially prepared, standardized tests or to develop its own local tests. Space limitations prevent a detailed discussion of the advantages and limitations of standardized testing; books are already written on this subject. Suffice it to say that standardized tests can be useful, and they provide helpful information (see Ebel [1979]). The community often wants to know how its youngsters compare with the normed group. Nonetheless, the teachers are also concerned that the test covers the content and processes they teach. They wish the tests to be more sensitive to instruction. Popham has observed that classroom teachers are beginning to become more comfortable in collaborating with test makers in the design and construction of tests that are "valid indicators of important skills" (Popham 1980).

It would seem that the school will benefit most from involving its teachers in the design and construction of the tests used in the assessment. Writing good test items and designing useful interviewing procedures is not a trivial matter. Mathematics educators and test specialists can give guidance and do most of the actual writing of test items, but the classroom teachers should be involved in determining whether or not the test measures mathematical skills and understandings that are clearly related to the way students learn them. Adequate time must be spent in formulating, trying out, and revising test items. It would not be unwarranted to take a full year to develop the measuring instruments. Time carefully spent here will pay off in better data later.

The classroom teacher's role in the process of instructing the test needs to be clearly understood. Teachers are full partners in the effort at assessment. One of their roles, especially in the elementary grades, will be to defend the child against unfair testing situations. Administrators, supervisory staff, and consultants need to recognize the advocacy role of the teacher and be sensitive to a teacher's criticism of a test item or testing procedure. Testing should be fair to the child. There is no sense in "tricking" the child, giving poor directions, or not allowing sufficient time to answer questions. However, an assessment is not a time for teaching. Teachers need to recognize the requirements for uniform procedures and the avoidance of clues that could jeopardize the integrity of the data. It is wise to spend time training the administrators of the test and developing careful directions and procedures for testing.

It is often believed that local teachers will make the tests too simple because of considerable pressure to have the school look good in the eyes of the community. It should be clear that no case can be made for building toward a high-quality program if the test gathers data only on minimal skills. Again, the purposes of the assessment should be reviewed. It may be wise to collect and analyze the data in a way that precludes the possibility of linking them to an individual student or to an individual classroom. The results could be viewed as a school-wide diagnostic profile of the needs and strengths of the mathematics program.

The committee should give careful thought to the population of students to be tested. Will every child be tested? A sample? Will learning disabled children be represented in the sample? Will their results be handled separately from the others? These and other similar questions need to be considered from many points of view, and the answers should be consistent with the major purposes of the assessment.

The committee must also consider the extent of the assessment in terms of the grade levels involved. It might be wise to block the assessment into manageable chunks. The first year could assess grades K–3 and 7–9; the second year could then involve grades 4–6 and 10–12. Successful techniques and clues learned during one phase of the assessment could be refined and incorporated during the second phase.

It is recommended that data be collected to allow for the observation of students' growth on specific objectives. Bright's chapter in the 1978 Yearbook of the NCTM shows how students' attainment of certain skills continues to grow following the grade levels in which the skill is usually considered "taught for mastery" (Bright 1978). The community will relate better to an assessment that shows, for example, that although only 60 percent of the third graders can subtract with renaming, by the sixth grade this result has climbed to 90 percent.

Two more aspects of data gathering deserve the committee's consideration. The first is the importance of good record-keeping procedures. Every attempt must be made to ensure the permanence, uniformity, and ready availability of the data gathered during the assessment. Sound and efficient procedures not only save time and money but also make the data more usable. Secondly, the committee should strive to avoid redundancy in testing. Some of the desired information may already exist from normally administered standardized tests, departmental final exams, statewide assessment, and so on. Classroom teachers are sensitive to the criticism of overtesting the students, and the committee would be wise to review all available sources of data and avoid duplication.

Interpretation of Results

There should be a rather obvious connection between the assessment and the interpretation of the results. Poor results in certain intrinsically difficult areas of content—fractions, for example—will not surprise many classroom teachers. Thus, it is not enough merely to report results. Some specific suggestions to help improve instruction or specific recommendations for possible curricular change must be given. The connection between the assessment and subsequent in-service programs or curricular modification is crucial. The principal benefit of objective-referenced tests is inherent in this need to have students' performance linked to classroom activity in specific ways.

Decisions on how to handle the data for the purpose of making interpretations must be made long before the data become available. The goals and objectives, the questions that are extensions of these goals, and the data-gathering instruments themselves must all fit together in concert to make possible relevant interpretation of results. The interpretation centers on the question, "What do these results tell us about our mathematics program?"

During the interpretational phase of the assessment the committee must review its consideration for protecting those desirable parts of the curriculum that were not tested. All instruction cannot be observed by the microscope of an assessment. Inferring needs and strengths only on what is tested to the disadvantage of what is not tested is dangerous. Some care taken here may prevent the erosion of nontested aspects of the program by ambitious plans resulting from the assessment results.

The discrepancy model is one strategy for interpreting the results of an assessment. One looks for discrepancies between expected results and actual outcomes. Thus, one of the requirements of the discrepancy model is to have realistic expectations for the results on specific objectives prior to testing. The actual precentage of students attaining an objective is relative to the intrinsic difficulty of the objective and to the timing of mastery of this objective in the present instructional sequence. For example, teachers would not expect students to perform well on the addition of unlike fractions when this topic is first introduced in the elementary school. A low result in and of itself would not be a concern. However, if there was a large discrepancy between their expectation and the actual results, then the teachers would be more disturbed and interested in possible explanations.

The assessment committee needs to be prepared to estimate how well the students will perform on each objective measured. This can be accomplished during the validation process. Experience has shown that mathematics educators and classroom teachers will vary widely on their predictions of students' performance. This does not necessarily render the discrepancy model unwieldy or unsatisfactory. It is necessary only that an attempt be made to discuss the expected outcomes and arrive at some general agreement about the approximate percentage of attainment. If the assessment is comprehensive and if some objectives are being tested at an early stage in the instructional sequence, then there will be low results. The committee will be looking for those instances in which the results are much lower or much higher than expected.

A successful assessment requires fast feedback of results and interpretations that are relevant to teachers. If, for example, the data are gathered in mid to late spring, then it is important to interpret the results before summer. Perhaps the teachers can obtain the instruments and school-wide results while they still have the students in their classes. The assessment may then be more relevant to the teachers. Workshops for discussing the interpretations could be held during the summer or early fall.

Summary

Further discussion on interpreting the results from an assessment, setting priorities, and extending outcomes from an assessment into plans for systematic change is given in subsequent chapters. We stop here for a summary of the major considerations involved in assessing the status quo.

A proper assessment is one in which the results are helpful in improving instruction and strengthening the mathematics program through curricular or instructional change. We start by examining preliminary data on issues that are relevant to the existing mathematics program. These issues are discussed relative to the forces that may influence certain decisions for change. The goals and objectives of a first-rate mathematics program are assembled. Important questions are posed regarding the attainment of specific goals. These questions shape the information-gathering effort. A feasible arrangement for the collection of data involving students' performance in the elementary school consists of the following:

- Timed tests of recall of the basic facts

- Paper-and-pencil, group-administered tests of computation
- Paper-and-pencil, group-administered tests of concepts and problem solving
- An individually administered interview designed to find out how the students think as they do their mathematics and to observe them perform with concrete materials in realistic situations

Data on students' achievement can be collected at the secondary level with an arrangement of tests and interviews similar to those used in the elementary school. Certain major objectives should be tested at more than one grade level so that growth is observable. Not every student at every grade level need be tested.

Data are also gathered from school records, the community, and follow-up studies of collegiate and vocational experience of graduates. Classroom teachers can be surveyed to determine the degree of current implementation of the program. This can be done as the goals, objectives, and test items are validated.

The discrepancy model can be used to identify needs and strengths. It is necessary to have worked closely with the classroom teachers to compare actual outcomes from the assessment with the teachers' expected outcomes. A survey of teaching practices and priorities can be conducted as a part of the assessment or as a part of the goal-setting process. The discrepancy model is not effective unless there has been some discussion of what to expect from the assessment.

The classroom teachers should be involved in the process of validating the goals and objectives from the beginning of the goal-setting process to the final gathering of data, interpretation, and priority-setting meetings. For the existing goals and objectives in the current program of instruction, the teachers can say whether or not the proposed measuring instruments and procedures parallel their instructional practices and emphases. For those areas represented by new goals, the teachers can testify to the current lack of instruction and to their desire to provide appropriate instruction as changes are formulated and new programs adopted.

The views of school administrators or the assessment committee leaders may differ with some teacher's view of when to teach fractions or whether geometry is important. As in many other areas of democratic decision making, fairness and openness should prevail in the assessment. It would be wise for the committee to be prepared to answer the questions that will inevitably be raised when test scores are lower than expected. Was the test of poor quality? Did the students misunderstand the questions? Or do teachers really try to teach this skill at this grade level? If the committee has collected data on the teachers' views of the extent to which current practices are related to the proposed tests, then teachers may have more genuine interest in the results of the assessment and exhibit more acceptance of the subsequent interpretations and recommendations that flow from the assessment venture.

We began this chapter with a question, "Why is it so difficult to conduct a proper assessment?" We have not answered the question directly, but the number of important considerations and caveats that have been raised certainly indicates the numerous ways in which an assessment can get off the track. School personnel seem to underestimate the importance of doing the job right. They often cut corners and undermine the effort at assessment; then they expect too much and are disappointed by the apparent inability to use the results to answer important questions. The failure of assessments to provide helpful information lies most often at the doorstep of poor planning, insufficient funding, and a lack of time for preparation and implementation. School leaders usually have a good track record when it comes to September-to-June projects. Ventures that require more than one year from conception to completion are often not carried out as well. A successful assessment of the status quo could well take two years to plan and implement.

REFERENCES

Begle, E. G. *Critical Variables in Mathematics Education: Findings from a Survey of the Empirical Literature*. Washington, D.C.: The Mathematical Association of America; Reston, Va.: National Council of Teachers of Mathematics, 1979.

Bright, George W. "Assessing the Development of Computation Skills." In *Developing Computational Skills*, 1978 Yearbook of the National Council of Teachers of Mathematics, edited by Marilyn N. Suydam, pp. 148–62. Reston, Va.: The Council, 1978.

Carpenter, Thomas P., Henry Kepner, Mary Kay Corbitt, Mary Montgomery Lindquist, and Robert E. Reys. "Results and Implications of the Second NAEP Mathematics Assessment: Elementary School." *Arithmetic Teacher* 27 (April 1980): 10–12, 44–47.

Coburn, Terrence G. "Statewide Assessment and Curriculum Planning: One State's Experience." *Arithmetic Teacher* 27 (November 1979): 14–20.

Conference Board of the Mathematical Sciences, National Advisory Committee on Mathematical Education (NACOME). *Overview and Analysis of School Mathematics, Grades K–12*. Reston, Va.: National Council of Teachers of Mathematics, 1975.

Ebel, Robert L. "Using Tests to Improve Learning." *Arithmetic Teacher* 27 (November 1979): 10–12.

Fey, James T. "Mathematics Teaching Today: Perspectives from Three National Surveys." *Arithmetic Teacher* 27 (October 1979): 10–14.

National Assessment of Educational Progress. *Mathematics Objectives Second Assessment 1978*. Denver, Colo.: Education Commission of the States, 1978.

National Council of Teachers of Mathematics. *An Agenda for Action: Recommendations for School Mathematics of the 1980s*. Reston, Va.: The Council, 1980.

_____. *A History of Mathematics Education in the United States and Canada*. Thirty-second Yearbook. Washington, D.C.: The Council, 1970.

_____. *Priorities in School Mathematics: Executive Summary of the PRISM Project.* Reston, Va.: The Council, 1981a.

_____. *Results from the Second Mathematics Assessment of the National Assessment of Educational Progress.* Reston, Va.: The Council, 1981b.

National Council of Supervisors of Mathematics. "Position Paper on Basic Skills." *Arithmetic Teacher* 25 (October 1977): 19–22.

Popham, W. James. "Educational Measurement for the Improvement of Instruction." *Phi Delta Kappan* 61 (April 1980): 531–34.

8

Designing a Plan for Change

THOMAS E. ROWAN
CATHERINE E. MORGAN

A LMOST everyone acknowledges the adage that change is inevitable. The only real question is whether we retain some kind of control over that change. Other chapters of this book have discussed the general nature of change, some kinds of resistance to change, and ways to determine a need and a direction for change. Sometimes change may be mandated by federal, state, or local governments. Minimum-competency testing has usually come about in this manner. Change may also be suggested by a national trend in society or education; the NCTM's publication *An Agenda for Action: Recommendations for School Mathematics of the 1980s* (1980) is an example. Often, the easiest changes to implement result from needs-assessment procedures in local schools or school systems.

The main purpose of this chapter is to discuss the design and implementation of plans to bring about a change after the decision has been made that a change is necessary for some reason. Basically, we shall focus on planning change on both a general and a specific level. On the general level, we shall describe characteristics of design and implementation. On the specific level, we shall present some models for change, both theoretical and real.

What Can Be Changed?

Changes can be made in any of a variety of dimensions of school programs. Some of these have greater implications for altering the total program than others (the domino effect). Some are easy to describe in the design phase but very difficult to accomplish in the implementation phase. Others are more difficult to design and somewhat easier to implement. Some possible areas for change and an elaboration of each are provided below.

Program rationale. The rationale of the mathematics program should always be developed at the local level. It should have input from as many of those who will use it as possible. Changes in the rationale should cause subsequent changes in other areas. By the same token, changes in other program areas should always be made within the parameters of the rationale. Too many people overlook the importance of a rationale in mathematics because they feel there is general agreement as to what is important. Although this feeling may be reasonably accurate, having no rationale makes a program vulnerable to frivolous changes that would otherwise not be defensible.

Curriculum. It has been stated that for many teachers the mathematics curriculum is primarily determined by the textbook that is used (ERIC 1980, p. 3). Standardized tests also determine at least a portion of the curriculum for many teachers, especially since the initiation of minimal-competency programs based on such tests. The curriculum *should* reflect the rationale of the program, the nature of mathematics, and a knowledge of how students learn. Texts and tests should be selected because they match the curriculum.

Instructional strategies. Recent research (Bloom 1976; Denham and Lieberman 1980) has supported a notion that many teachers believe but often feel unable to use—students learn better when the time they spend on appropriate tasks is increased. What this means is that the material being taught should be based on the needs of the students and *not* on the textbook chapter or any other preestablished content pattern. Methods of instruction should reflect the latest knowledge of how students learn at various age levels.

Information collection and management in the classroom. To some extent, this area is a part of the previous one; it has been separated for the purpose of emphasis. If instruction is to be based on the readiness of students—on a good knowledge of their current achievement—then some reasonable and systematic means for collecting and organizing information about students' achievement is essential. Every effort must be made to do this in a manner that does not demand too much of the teacher's scarce instructional time.

Evaluation. School systems should have some kind of agreement about how they will measure their success. This agreement needs to involve administrators, teachers, parents, the community, and, when appropriate, students. Too often, when no common agreement has been reached, success or failure is determined by politicians or news reporters.

Models for Designing and Implementing Change

The notion of using a model to help plan change has several purposes. A model is often associated with efficiency because the model enables us quickly to see and pursue the steps suggested by its developer, usually with some assurance that it has been tried before. Furthermore, a model may suggest some important activities that might otherwise be overlooked. We can always *adapt* a model to our local circumstances, but it gives us a starting place from which we can move forward efficiently.

The design and implementation of changes in any of the areas mentioned require careful planning. Sometimes externally developed programs are brought in for local adoption or adaptation. At other times, a program may be initiated and developed entirely as a local effort.

Critical differences in planning exist between changes that require developmental work and those that do not. An illustration showing a general model with various paths to installing and implementing a new program is shown in figure 8.1 (Sikorski et al. 1976). Treat the major headings shown across the top of this figure as major steps in a path leading to change.

Two terms in figure 8.1 may require explanation. The term *invention* is used to indicate who designs and develops the materials to be implemented in response to the need. The term *fidelity* is used to indicate whether the materials are used as designed by the inventor or whether they are modified locally to fit circumstances.

When externally developed material is adopted locally, the major tasks confronting the school system are in the areas of local dissemination, inservice training, and program evaluation. Often an external innovation includes a training package or model.

First consider the simpler situation—that of the adoption of an externally developed package (see fig. 8.2). Basically, the use of such a package requires communication, training, and evaluation. These general areas then break down into more specific components, such as "evaluation of program" (6.1).

To illustrate the design and implementation of a plan that adopts an externally developed program, we shall describe an example of a real situation using each step in figure 8.2. This situation occurred at the school level; it could have been developed as easily at a system-wide level.

Adopting an External Program at a School Level

Perceived need (1). A group of concerned teachers from a local school contacted their mathematics supervisor. They felt that their mathematics program had become too paper-and-pencil oriented and that the pupils were not experiencing good conceptual development. A meeting was set up with the entire faculty, and the situation was discussed. Although some disagreement arose about the real existence of a problem, there was general agreement that efforts to improve the mathematics program would be acceptable. It was also agreed that the efforts for improvement should focus primarily on concept development, since computational skills seemed to be progressing

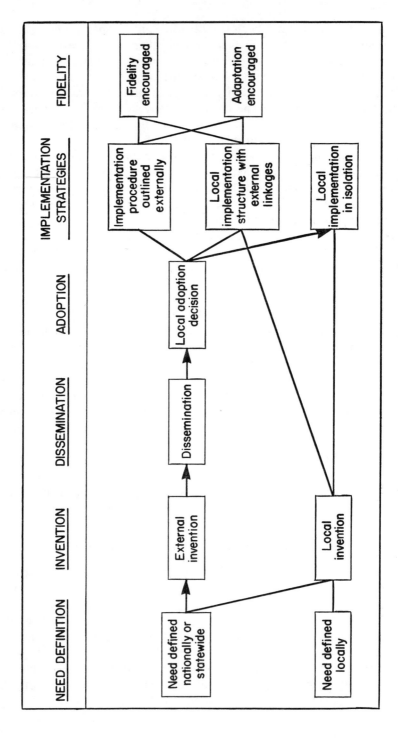

Fig. 8.1. Alternative paths toward innovation. (Redrawn from Sikorski et al. 1976.)

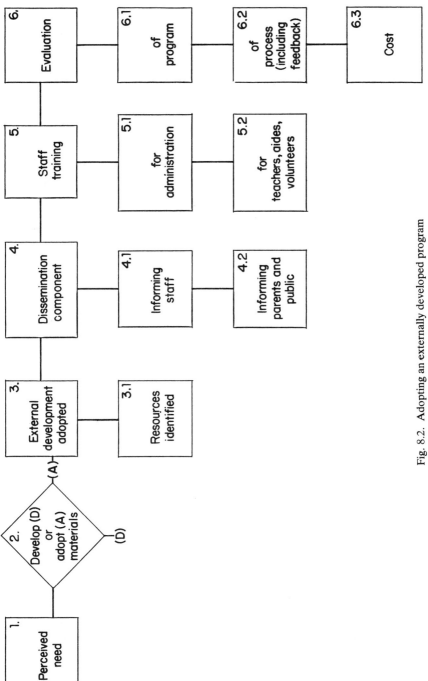

Fig. 8.2. Adopting an externally developed program

reasonably well. A second meeting was planned at which each teacher was invited to suggest whatever attack on the problem he or she felt was appropriate.

Develop or adopt materials (2). The second meeting enabled the teachers and the supervisor to discuss the possibility of making changes through the adoption of a locally developed program versus an externally developed program. The majority of the teachers noted that local activities in response to a problem of this sort would be very difficult at the school level. They also noted that several externally developed programs claimed to have made good progress in meeting needs of this sort. If one of these federally funded or commercial programs could meet the need of the school, then that would appear to be the more efficient approach to the problem.

External development adopted (3). After a review of several programs, a particular program and its materials were identified as appealing to the teachers, as having research evidence that appeared to support their effectiveness, and as being readily available for use at the beginning of the next school year.

Resources identified (3.1). Two kinds of resources were considered, financial and human. The school had some money for texts and instructional materials but was reluctant to put it all into mathematics while neglecting the needs of other disciplines. An appeal was made to the school system through the mathematics supervisor. Could some school district funds be found to help? This school system had wisely set aside some funds to encourage promising local initiatives. This money was offered to the school on a matching basis and was accepted. The human resources needs were of three types: leadership to provide initial training and mathematics expertise, ongoing leadership and coordination of the in-school program, and teachers willing to participate enthusiastically. The local mathematics supervisor agreed to supply the initial leadership, a teacher within the school was willing to do the in-school coordination, and the other teachers agreed to use faculty meetings during the early part of the upcoming year to receive in-service training. Had this been a secondary school, a department chairperson might have taken on the role of the in-school coordinator.

Dissemination component (4, 4.1). In this particular project the dissemination of information to the staff had already occurred in the process of discussing the problem and selecting the program. No further efforts to inform the staff were required. If this had been a project involving the adoption of an external innovation at the system-wide level, then more extensive attention to informing the staff would have been necessary.

Informing parents and public (4.2). It should be pointed out that at all stages of the previous discussion the school principal had been extensively involved as the instructional leader of the school. Since this was a school project, it was particularly important to inform parents early. A PTA meeting was held in the spring of the school year preceding the initiation of the project at the classroom level. At this meeting the principal, the mathematics supervisor, and several of the teachers described the program and responded to questions. Had this been a project of the entire school system, a number of other public relations possibilities might have been considered. These are illustrated in figure 8.3.

Major Task: Informing Parents and Public

Subtasks	Responsibility	Time
Newspaper announcements of adoption	Local projector initiator, superintendent, or information department	Immediately following school board action
Meetings of PTA executive boards in involved schools	Principals	Early September
Announcement in PTA newsletter	PTA editor	October
PTA meeting	PTA president, principal, and teachers	November
Follow-up information to parents	Principal	As needed
Classroom demonstration	School staff	February
Information on evaluation report	Department of research or similar office	May–October

Fig. 8.3

Staff training (5). Because the key to success in the change process lies with the school staff, careful planning of in-service training is of paramount importance. One example of a planning guide for staff training is shown in figure 8.4. The use of such guidelines can stimulate questions in the mind of the planner of the program as well as keep the district administrators informed about details that are useful in managing and in future planning.

Training for administration (5.1). Since the adoption of the new program was initiated in the school, the administration was involved from the onset and needed no additional training. Had this been a system-wide effort, some special provision for administrative training would have been essential. There is mounting evidence that support and commitment by the school's administrators is a critical factor, perhaps the *most* critical factor, in the

Planning Guide for Staff Training

1. Name of planner _____

2. Name of program _____

3. Date _____ 4. Length _____

5. Objectives of program

_____

_____

_____

_____

6. Target audience (include level) _____

7. Location _____

8. Evidence of need _____

_____

_____

_____

_____

9. Evaluation plan _____

_____

_____

_____

10. Previous course number (for repeated offerings) _____

    Has there been a modification? Yes _____ No _____

    Previous date _____

11. Name of leader_____

12. Type of program (check one or more)

    Laboratory_____   Student demonstration _____

    Regular class _____   Readings _____

    Other (please state) _____

13. Need and purpose for funds _____

    _____

Fig. 8.4

success of a project. Training for administrators need not be the same as that for teachers, but the administrators must be involved and must support the activities of the in-service program. There must be a direct focus on what the principal should look for and do to contribute to the success of the project.

Training for teachers (5.2). In this adoption by a local school, teachers' training consisted of two phases—training the in-school coordinator and training the other teachers. The publisher of the program offered summer leadership workshops in several locations at no cost other than travel expenses. The coordinator and the mathematics supervisor went to one of these summer workshops. The training for the other teachers began in the fall before the children returned to school. Teachers were normally given four paid days before the students reported so that they could plan and set up for the school year. Two half-days of these four days were used to do the initial training of teachers. The sessions were conducted by the principal, the coordinator, and the mathematics supervisor. During the first half of the school year, teachers met four times by grade-level teams and twice as a total faculty to continue their training and to discuss problems and successes. During the first year the teachers participated in the training without the benefit of external incentives. During subsequent years the program was changed to an in-service credit course, and the credit could be applied toward salary improvements. Incentives to encourage wider and more regular participation are often used as a part of the teacher-training activity. Such incentives can include credits, stipends, release time, new teaching materials, and other encouragements.

Evaluation (6). This phase of the change process is discussed in chapter 10 and will not be pursued here.

Developing a Local Program

If externally available materials do not appear to meet the needs that have been identified, then local development may be the appropriate response. Figure 8.5 starts at the same point as figure 8.4 and pursues the development course of action. The major part of this course is the actual development process. Once this stage is completed, the other stages are similar to those in external adoption.

The locally developed project described here is called the *Instructional System in Mathematics* (ISM) and was carried out by a large public school system. It was an extensive undertaking and the process may therefore seem overly complex, but the basic steps and techniques could be applied to any project.

Perceived need (1). The school system perceived that little relevant infor-

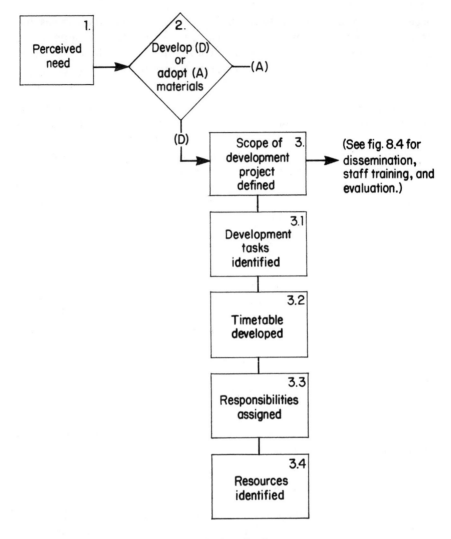

Fig. 8.5. Developing a local program

mation existed on which decisions could be made on how to improve instructional programs. The school board was concerned about basic skills. Standardized test results do not provide much information; these test results are intended to compare the performance of a local group with national performance. Furthermore, only a portion of the curriculum can be tested in a multiple-choice format, and data on many important outcomes of the school program are often not available.

Develop materials (2). For the reasons stated above, the decision was made to develop K–8 instructional systems in the four basic skills areas— mathematics, reading and language arts, science, and social studies. Inasmuch as mathematics had the most advanced curriculum with the K–8 objectives already well defined, the decision was made to develop the Instructional System in Mathematics (ISM) first.

Scope of development project defined (3). Very early in the planning it was decided that initial development should focus on providing information that could be used by the teacher to improve instruction at the classroom level. It was further decided that the project should encompass the total K–8 mathematics program. The components of the system were listed: mathematics objectives (curriculum), placement, instruction, assessment, and reports. In-service training and evaluation were treated as necessary functions but were not considered parts of the system. The decision makers discussed several phasing-in possibilities. The development could occur in stages; for example, the curriculum could be developed one year, the assessment the following year, and the reporting function at a later time. Alternatively, the complete development could be undertaken the first year in pilot schools and other schools phased in over several years. Because of the desire to have information reported as soon as possible, the latter course of action was adopted. It was agreed that the development of the total program in a relatively short period of time (six months) would make it essential that the first year of implementation in schools be treated as a developmental year with schools serving as user-developers.

This plan required an intensive developmental effort with the involvement of many different departments and individuals within the school system. To assure the necessary commitment of support, an Instructional Systems Development Project (ISDP) was formed from the existing departments. It included staff members from Computer Related Instruction (CRI) and the Department of Curriculum and Instruction (DCI) and mathematics specialists from area offices that worked directly with the schools. CRI representatives included design specialists and computer programmers; DCI representatives included mathematics specialists, and the area staff brought mathematical and implementation expertise. Since both assessing and reporting were tentatively planned to involve the use of the central computer, regular and close contact was required with the Department of Management Information and Computer Services (DMICS).

It would again seem important to stress a cautionary note: this description is of a complex project in a large school system. Although a smaller project or a similar project in a smaller school system would obviously be different, the principle of the description above would remain the same—*it is vital to include as many concerned individuals or offices as possible and to do so as early as possible*. Commitment and cooperation are not achieved by surprising an affected individual with a requirement after it is too late to make any modifications.

Development tasks identified (3.1). ISDP had the responsibility for designing and implementing a mathematics system in a very short period of time. Careful planning was essential if this effort was to have any hope of success. The process used to accomplish this was simple but effective. The total project was given the brainstorming treatment. Major components of the project were defined and subtasks under these were constructed. The components selected were as follows: pilot schools, information sessions, objectives/test items/mastery/standards, assessment, instructional resources, reports, and management. These components form the headings in figure 8.6.

INSTRUCTIONAL SYSTEMS DEVELOPMENT (MATHEMATICS)

Tasks, Groups Responsible, and Dates of Completion

Event No.	Task	Responsibility	Completion Date
	Pilot Schools		
A-1	Select and notify pilot schools	Area asst. supts. Deputy supt.	3/31
A-2	Plan and conduct first half-day orientation for staff	DCI Math ISDP	5/1
A-3	Plan and conduct second half-day orientation for staff	DCI Math ISDP	6/2
A-4	Set objectives, prepare materials, and plan one-week in-service course for staff	Area math teacher specialists DCI Math ISDP	8/1
A-5	Teach in-service course for staff	"	8/27
A-6	Provide regular support during the school year	"	6/1
	Information Sessions		
B-1	Identify initial groups to be informed and set dates	Area directors for instruction Deputy supt.	3/31

Event No.	Task	Responsibility	Completion Date
B-2	Develop script and slide presentation	DCI Math ISDP Educational Materials Dept. (DEMAT)	4/15
B-3	Make presentations	DCI Math ISDP	6/15

Objectives, Test Items, Mastery, Standards

Event No.	Task	Responsibility	Completion Date
C-1	Designate topics for students' profiles at each grade level	DCI Math	3/15
C-2	Identify key objectives and coding for all objectives	DCI Math	3/15
C-3	Expand objectives to include standard and mastery levels	DCI Math ISDP	4/15
C-4	Prepare sets of assessment items for key objectives	DCI Math Area teacher Specialists ISDP	4/15
C-5	Develop hierarachies for each topic	ISDP	5/1
C-6	Prepare clusters of interrelated objectives for use by students and teachers	ISDP	5/1

Assessment

Event No.	Task	Responsibility	Completion Date
D-1	Flowchart and code diagnostic sequences for CAI schools	ISDP	8/1
D-2	Prepare diagnostic tests for non-CAI schools	DCI Math ISDP	7/1
D-3	Flowchart and code sets of assessment items for CAI schools	ISDP	8/1
D-4	Print diagnostic tests and sets of assessment items for non-CAI schools	DEMAT	8/15
D-5	Set specifications for op-scan use with assessment items for non-CAI schools	DMICS ISDP	4/15
D-6	Develop program for transfer of data from op-scan sheets to pupils' master files	DMICS ISDP	8/1

Event No.	Task	Responsibility	Completion Date
D-7	Code on-line course for entering K–2 data in CAI schools	ISDP	8/1
D-8	Design and develop programs to transfer students' profiles to pupils' master files	ISDP DMICS	8/1

Instructional Resources

Event No.	Task	Responsibility	Completion Date
E-1	Design format for instructional guide	DCI Math	6/15
E-2	Prepare guide sheet(s) related to each objective	Summer assigned teachers DCI	8/1
E-3	Print instructional guides	DEMAT	8/15
E-4	Prepare form for inventory of classroom and school instructional materials and descriptions of organization for instruction	ISDP DEMAT	5/1
E-5	Inventory approved materials at pilot schools	Media specialists Teachers/pilot schools	6/15
E-6	Provide additional instructional materials to pilot schools where need is indicated	DCI Math	9/1

Reports

Event No.	Task	Responsibility	Completion Date
F-1	Design a format for several possible profile and progress reports for classes, groups, and schools	ISDP DMICS	5/1
F-2	Secure approval from school staffs	Staffs/pilot schools	5/15
F-3	Develop programs for school reports	ISDP DMICS	10/1
F-4	Design a format for several possible profile and progress reports for superintendent, BOE, and areas	ISDP DMICS	5/1
F-5	Secure approval from group involved	Involved group	5/15
F-6	Develop program for area- and system-wide reports	ISDP DMICS	10/1
F-7	Develop specifications for validation reports	ISDP DCI Math	8/1

Event No.	Task	Responsibility	Completion Date
F-8	Develop programs for validation	ISDP	1/1
F-9	Upgrade CAI utility reports for ISM	ISDP	10/1
F-10	Test non-CAI sample for reliability	Research Dept. (DRE)	2/1

Management

G-1	Set performance targets	DRE DCI	8/1
G-2	Prepare orientation for students in pilot schools	Summer assigned teachers	8/1
G-3	Orient students to system	Teachers/pilot schools	9/30
G-4	Report to executive staff on system development	Deputy superintendent	Monthly beginning 4/1
G-5	Monitor and evaluate progress of students	Staffs/pilot schools	Weekly beginning 10/1
G-6	Monitor and evaluate performance of system	DCI Math	At regular intervals beginning 1/1
G-7	Monitor and evaluate efficiency of system	ISDP	Monthly beginning 10/1
G-8	Analyze components regarding students' achievement	Area staff DCI Math	Monthly beginning 11/1
G-9	Modify components	DCI Math ISDP	As required

Fig. 8.6

Timetable developed, responsibilities assigned (3.2, 3.3). Some idiosyncrasies of this particular project create unique subtasks in this process (e.g., "key objectives" and "coding" were needed to facilitate computer storage of students' records), but again the usefulness of this type of organization for the planning process should be evident. Once this list was completed to the point where the ISDP staff felt that most essential steps were covered, then responsible offices were identified and necessary (or desirable) completion dates were established. This chart was useful in this form, but it was made even more useful by its transformation into the form seen in figure 8.7. This actually becomes a month-by-month work plan for the developmental effort. It is vital that this particular timetable be used with a "looking ahead" approach (e.g., it is essential to look at E-3 in August to see that the instructional guides must be designed and developed primarily during June, July, *and* August). That is to say, for this particularly complex task, separate subplans are required. The subplan for the development of instructional guides is shown in figure 8.8. Again, the emphasis is on the process and not on the particular project. It is again important to cite the need for full cooperation among various offices or individuals. Some teachers and area staff were employed in the summer to develop the guide. The print shop devoted most of its energies over several weeks to printing and assembling the guide. The teachers in twelve elementary schools and three junior high schools were paid to come back to work three days early so they could learn about the guide and the management of the program.

Resources identified (3.4). Once the plan for the project has been completed, the process of identifying resources begins. As responsibilities are assigned, human and other resources are thereby being identified. Additional resource requirements that do not show up in this process (e.g., funds to pay for the early return of teachers) must be culled from the nature of the task itself.

INSTRUCTIONAL SYSTEMS DEVELOPMENT MILESTONES

Completed by End of		Tasks	Groups Responsible
March	C-1	Designate topics for students' profiles	DCI Math ISDP
	C-2	Identify key objectives and coding for all objectives	DCI Math ISDP
	A-1	Select and notify pilot schools	Area asst. supts. Deputy superintendent
	B-1	Identify initial groups to be informed and set dates	Area directors for instruction Deputy Supt.
April	G-4	Report to executive staff on progress of systems development	Deputy Supt.
	B-2	Develop script and slide presentations	DCI Math ISDP DEMAT

Completed by End of		Tasks	Groups Responsible
	C-3	Expand objectives to include standards and mastery	DCI Math ISDP
	C-4	Prepare sets of assessment items for key objectives	DCI Math Area teacher specialists ISDP
	D-5	Set specifications for op-scan use with assessment items for non-CAI schools	DMICS ISDP
	A-2	Plan and conduct first half-day orientation for school staffs	DCI Math ISDP
	C-5	Develop hierarchies for each topic	ISDP
	C-6	Prepare clusters of interrelated objectives for use by students and tearchers	ISDP
	E-4	Prepare form for inventory of classroom and school instructional materials and description of organization for instruction	ISDP DEMAT
	F-1	Design format for several possible profile and progress reports for classes, groups, and schools	ISDP DMICS
	F-4	Design format for several possible profile and progress reports for superintendent, board of education, and areas	ISDP DMICS
May	G-4	Report to executive staff on progress of systems development	Deputy supt.
	F-2	Secure approval for report formats from school staffs	Staffs/pilot schools
	F-5	Secure approval for report formats for areas, board of education, and superintendent	Involved group
June	G-4	Report to executive staff	Deputy supt.
	A-3	Plan and conduct second half-day orientations for school staffs	DCI Math ISDP

Completed by End of		Tasks	Groups Responsible
	B-3	Make presentations to identified groups	DCI Math ISDP
	E-1	Design format for instructional guide	DCI Math ISDP
	E-5	Inventory approved materials at pilot schools	Media specialists Teachers/pilot schools
July	G-4	Report to executive staff	Deputy supt.
	D-2	Prepare diagnostic tests for non-CAI schools	DCI Math ISDP
August	G-4	Report to executive staff	Deputy supt.
	A-4	Plan one-week in-service course for school staffs	Area math teacher specialists DCI Math ISDP
	D-1	Flowchart and code diagnostic sequences for CAI schools	ISDP
	D-3	Flowchart and code sets of assessment items for CAI schools	ISDP
	D-6	Develop programs for transfer of data from op-scan sheets	DMICS ISDP
	D-7	Code on-line course for entering K–2 data on CAI schools	ISDP
	D-8	Design and develop programs to transfer students' profiles to pupils' master files	ISDP DMICS
	E-2	Prepare guide sheet(s) related to each objective	Summer assigned teachers DCI
	F-7	Develop specifications for validation reports	ISDP DCI Math
	G-1	Set performance targets	DRE DCI
	G-2	Prepare orientation for students in pilot schools	Summer assigned teachers
	D-4	Print diagnostic tests and sets of assessment items for non-CAI schools	DEMAT
	E-3	Print instructional guides	DEMAT

Completed by End of		Tasks	Groups Responsible
	A-5	Teach in-service course for school staffs	Area math teacher specialists DCI Math ISDP
September	G-4	Report to executive staff	Deputy supt.
	E-6	Provide additional instructional materials to pilot schools	DCI Math
	G-3	Orient students to system	Teachers/pilot schools
	F-3	Develop programs for school reports	ISDP DMICS
	F-6	Develop programs for area- and system-wide reports	ISDP DMICS
October–June			
	G-4	Report to executive staff	Deputy supt.
	A-6	Provide regular support to pilot schools	Area math teacher specialists DCI Math ISDP
	F-8	Develop validation programs	ISDP
	F-9	Upgrade CAI utility reports for ISM	ISDP
	G-5	Monitor and evaluate progress of students	Staffs/pilot schools
	G-6	Monitor and evaluate perform- ance of system	DCI Math
	G-7	Monitor and evaluate efficiency of system	ISDP
	G-8	Analyze components regard- ing student achievement	Area staff DCI Math
	G-9	Modify components	DCI Math ISDP

Fig. 8.7

Item E (see fig. 8.6), the instructional guide, is possibly the most difficult of all the tools to develop. The difficulty arises from the complexity of the project. The instructional guide should include recommended texts, materials, and equipment *keyed to* the objectives. It should also contain sample activities, developmental sequences for major objectives, enrichment objectives and activities, and appropriate suggestions of methods for teaching basic skills. It may be acceptable to develop parts of the guide after the pilot year is in progress, but the keying of texts, materials, and equipment must be completed before that. It is estimated that this very difficult task will probably take all available time if it is to be completed prior to the pilot year. The following calendar (fig. 8.8) separates the task into two parts, the keying activity and the other activities.

Product 1: A list of the mathematics objectives with recommended materials, texts, and equipment keyed to each objective

Date needed: 2 August

27 work days: 1 day each week, 1 March to 11 June
10 days, 7 July to 21 July

Completed draft: 21 July
Typing: Concurrent with drafting
Graphics: Concurrent with drafting
Printing: 19 July to 30 July
Completion: 2 August

Staff required: 7 writers (2 teacher specialists, 2 elementary teachers, 1 junior high school resource teacher, 1 junior high school teacher, 1 media specialist)

Tasks to be completed:
1. Decide on the format and physical structure of the guide, keeping in mind that it must be updated annually.
2. Survey selected MCPS elementary and junior high school teachers for recommendations of materials, equipment, and perhaps text pages for specific objectives.
3. Update MCPS list of manipulative materials.
4. Compile a list of nontext printed materials.
5. Update the list of films, filmstrips, and transparencies.
6. Compile a list of texts and text materials (DMP, IMS, etc.).
7. Key all materials to objectives.
8. Work on sample activities, developmental sequences, and so on as time permits.

Product 2: Sample activities, developmental sequences for major objectives, enrichment objectives and activities, introductory and retention levels, and appropriate suggestions of methods for teaching basic skills

---

Date needed: 1 August (It would be highly desirable to have all or part of this material for the pilot year.)

Begin: 30 June (Include regular meetings during the school year and two weeks for summer workshop each summer.)

Complete draft: Mid-July

Printing completed: 1 August

Staff required: 7 writers (same as for keying activity)

Fig. 8.8

The steps outlined above have focused primarily on the planning events for internal *development*. Some training was mentioned, but the full development of dissemination, training, and evaluation has been omitted, since they would be very similar to the parallel steps for adopting an external program, and these were described earlier (see fig. 8.4).

The project described above was quite complex and took place in a large school system. It may seem to have been presented in unnecessary detail, but the specific steps that are included make clear the extensive interaction and involvement and also may suggest details that might otherwise be overlooked. On a smaller scale, the *general* nature and purpose of the interactions and steps would remain the same.

Workshops for Planning Change

One specific way to approach the planning process has been suggested by Burton (1980). Figures 8.9, 8.10, and 8.11 describe a comprehensive planning workshop and two approaches to carrying out the planning process, one that uses three days and another that uses a half-day session. A very high level of readiness would doubtless be necessary to complete the tasks within the half-day session. Some persons might be more comfortable with even more than three days of planning. The workshop agenda provides some detail on the kinds of activities used in applying this particular model to achieve a plan. This process could then lead to the adoption of an external program or the development of an internal program.

Summary

Change can be easy or difficult, effective or ineffective, efficient or inefficient. To a large extent, the quality and quantity of change that occurs depends on the quality and quantity of the planning that precedes it. This chapter has presented some general characteristics to be considered in

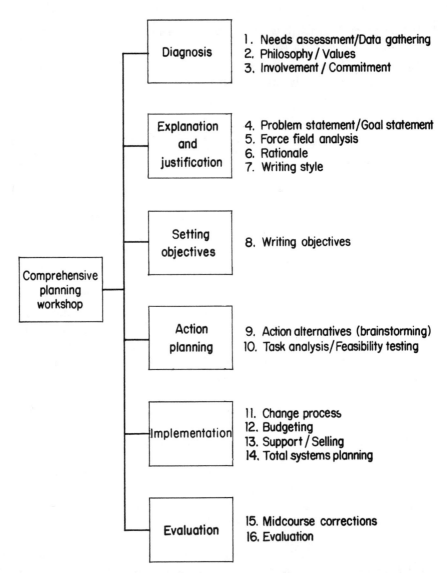

Fig. 8.9. Overview of comprehensive planning workshop. (Prepared by Charles Schwahn for the South Dakota Department of Education and based on materials by Kenneth Blanchard.)

planning change as well as some models that may be useful in the planning process. There is no simple, prepackaged scheme that will guarantee success in making lasting and worthwhile changes. Using some kind of systematic planning process is certainly valuable in maximizing the probability of success.

GAC Planning Workshop

Monday, 29 March	Tuesday, 30 March	Wednesday, 31 March
1:00 p.m.—Planners and operators game	8:30 a.m.—Force field analysis Handout/Discussion/Individual or small group work	8:30 a.m.—Budgeting Handout/Discussion
2:30 p.m.—Introduce workshop—structure and expectations	9:15 a.m.—Rationale/Justification of need Handout/Individual work	9:15 a.m.—Support-Selling/Strategies for gaining acceptance of plan Handout/Discussion
3:00 p.m.—Needs assessment in planning Handout/Individual work	9:45 a.m.—Break	9:45 a.m.—Systems planning Handout/Discussion
3:45 p.m.—Philosophy/Values Discussion	10:15 a.m.—Writing objectives Handout/Individual work	10:00 a.m.—Break
4:00 p.m.—Involvement/Commitment Handout/Individual work	11:15 a.m.—Strategies for prioritizing objectives	10:15 a.m.—Midcourse corrections Handout/Discussion
4:20 p.m.—Problem/Goal statement Handout/Individual work	12:00 noon—Lunch	10:40 a.m.—Evaluation Handout/Discussion
4:45 p.m.—Achieving goal consensus ISE-We agree process	1:15 p.m.—Strategies for generating proposed activities/Alternatives • Brainstorming • Force field analysis • Reviewing research • Clearinghouse for models	11:00 a.m.—Characteristics of effective helping relationships/Summarize paraphrasing-listening skills
	2:30 p.m.—Task analysis/Feasibility analysis/Establishing timelines Handout/Individual work	
	3:15 p.m.—How change occurs in systems/A model for change Game/Handout/Discussion	

Fig. 8.10. (Prepared by Charles Schwahn for the South Dakota Department of Education and based on materials by Kenneth Blanchard.)

Planning Workshop Involving Parents

1:30 p.m.—Introduction
1:45 p.m.—Developing a problem statement/Goal statement
2:15 p.m.—Force field analysis
2:45 p.m.—Brainstorming/Writing objectives
3:00 p.m.—Break
3:15 p.m.—Generating proposed activities/Task analysis/Gaining support
4:15 p.m.—Implementing the action plan
4:30 p.m.—Closing

Fig. 8.11

Some vital ingredients for implementing change that have been identified in this chapter are the following:

- Relating needs to resources and deciding whether to *develop or adopt* a program to bring about a change
- *Identifying needed resources* for an adopted program
- *Defining the scope* of a locally developed program
- *Identifying developmental tasks*
- *Developing a timetable*
- *Assigning responsibilities*
- *Identifying needed resources* for a locally developed program
- *Disseminating information* to staff, parents, and the public
- *Training* administrators, teachers, aides, and volunteers
- *Evaluating* the program, the process, and the cost

REFERENCES

Bentley, Ernest L., Jr. "Comprehensive Staff Development Model." *Theory into Practice* 11 (October 1972): 262–66.

Bloom, Benjamin S. *Human Characteristics and School Learning.* New York: McGraw-Hill Book Co., 1976.

Burton, Grace M. *Comprehensive Planning Workshop Overview.* Chapel Hill, N.C.: University of North Carolina, 1980.

Denham, Carolyn, and Ann Lieberman, eds. *Time to Learn.* Washington, D.C.: U.S. Department of Education, National Institute of Education, 1980.

ERIC Clearinghouse for Science, Mathematics and Environmental Education. *Summary of NSF Literature Review in Mathematics Education.* Information Bulletin, Summer 1980. Columbus, Ohio: ERIC/SMEAC, 1980.

Jung, Charles, and Ronald Lippett. "The Study of Change as a Concept—in Research Utilization." *Theory into Practice* 5 (February 1966): 25–29.

National Council of Teachers of Mathematics. *An Agenda for Action: Recommendations for School Mathematics of the 1980s*. Reston, Va.: The Council, 1980.

Sikorski, Linda A., Brenda J. Turnbull, Lorraine I. Thorn, and Samuel R. Bell. *Factors Influencing School Change: Final Report*. San Francisco, Calif.: Far West Laboratory for Educational Research and Development, 1976.

9

Developing Support for Change

ROSS TAYLOR

C HANGE in the curriculum is made by people, and support for effective change by all concerned persons is desirable. Support by the teachers and principals involved is essential. The efforts of these key individuals who assume leadership roles can make the difference between the success or failure of an attempted change. This chapter will present practical ideas for those leaders to develop support for implementing the NCTM's *Agenda for Action: Recommendations for School Mathematics of the 1980s* (1980) in the schools.

Climate for Change

Implementing curricular changes is much easier in a school situation where the climate for change is positive. In such a climate, change is encouraged and productive change is positively reinforced. Faculty members should feel that they can take risks. If they feel that they must be 99.9 percent sure that the change will be successful, they will never try anything very bold.

In a favorable climate for change, faculty members are encouraged to try new ideas. Those changes with positive effects are identified and replicated through appropriate evaluative procedures. The changes with negative effects are quickly identified and either modified or discarded. Such procedures maximize the beneficial effects of change, minimize the negative effects, and contribute to a positive climate for change. Risks can be taken because the potential payoffs are high and the potential detrimental effects are low. The gains will far outweigh any losses.

The administration often plays a key role in creating a climate for change. However, even when the administration does not assume leadership for creating such a climate, teachers and persons in positions of leadership in curriculum development can take the initiative. Teachers can work together in a professional way and provide each other with positive reinforcement in efforts for change. Department heads, supervisors, and college mathematics educators can also stimulate change and provide positive reinforcement.

Just as a field needs to be made ready for planting, a school environment needs to be readied for change. It makes no sense to plant seeds in concrete. Positive leadership can go a long way in providing a setting in which change can thrive.

Leadership for Change

Leadership and Power

For change in the mathematics curriculum to occur, someone must exert leadership. School district administrators and school principals can exert leadership by creating a climate for change, by stimulating curriculum leadership within the professional staff, and by seeking leadership from external sources. The leadership for specific changes in the mathematics curriculum often must come from the mathematics education community. Professionally active teachers and supervisors as well as mathematics educators external to the school system can provide this type of leadership.

Leadership is the process of attempting to influence others, and power is the means by which compliance is gained. Persons who wish to exert leadership in mathematics education should become familiar with the kinds of power associated with leadership. Hersey, Blanchard, and Natemeyer (1979) have identified the seven bases of power listed below (p. 1):

COERCIVE POWER is based on fear. A leader high in coercive power is seen as inducing compliance because failure to comply will lead to punishment such as undesirable work assignments, reprimands, or dismissal.

CONNECTION POWER is based on the leader's "connections" with influential or important persons inside or outside the organization. A leader high in connection power induces compliance from others because they aim at gaining the favor or avoiding the disfavor of the powerful connection.

EXPERT POWER is based on the leader's possession of expertise, skill, and knowledge, which, through respect, influence others. A leader high in expert power is seen as possessing the expertise to facilitate the work behavior of others. This respect leads to compliance with the leader's wishes.

INFORMATION POWER is based on the leader's possession of or access to information that is perceived as valuable to others. This power base influences others because they need this information or want to be "in on things."

LEGITIMATE POWER is based on the position held by the leader. Normally, the higher the position, the higher the legitimate power tends to be. A leader high in legitimate power induces compliance or influences others because they feel that this person has the right, by virtue of position in the organization, to expect that suggestions will be followed.

REFERENT POWER is based on the leader's personal traits. A leader high in referent power is generally liked and admired by others because of personality. This liking for, admiration for, and identification with the leader influences others.

REWARD POWER is based on the leader's ability to provide rewards for other people. They believe that their compliance will lead to gaining positive incentives such as pay, promotion, or recognition.

Leaders for change in the mathematics curriculum are usually not in positions that have high coercive power. Their reward power is usually limited only to the power to give recognition. They sometimes also have legitimate power or connection power. Consequently, they need to draw more on expert power, information power, and referent—or personal—power. In other words, they must use their expertise, their knowledge, and their personality to influence change. Their influence can be extended if they help others with whom they work in the mathematics education community to develop their capacity for leadership.

Developing Leadership for Change

As experienced leaders work on developing leadership in others, they should involve them in activities in which they will be successful. At first they should participate in relatively small, short-term professional activities. As they accomplish these activities successfully and receive praise for their accomplishments, they will be ready and willing to tackle more demanding activities. In the process, they will become a positive resource and force for change. However, if they are allowed to bite off more than they can chew and find themselves in a situation of failure, they will tend to turn off and perhaps become a force that opposes change.

The NCTM and its affiliated groups offer excellent opportunities for experiences in leadership. Teachers who gain experience through meetings, publications, committees, and offices prepare themselves to assume greater responsibilities in their school systems. In recent years in Minneapolis, more teachers have gone into administration from mathematics than from any other field. Many of these mathematics teachers developed their leadership skills through local curriculum activities and through activities of the Minneapolis Mathematics Club, the Minnesota Council of Teachers of Mathematics, and the NCTM. This phenomenon has led to the coining of the term "the mathematics mafia." Nevertheless, these former mathematics teachers form an important support structure within the administration for change in mathematics.

Developing and stimulating leadership contributes to a positive climate for change. Change will become easier with the broadest possible development of leadership that will then serve as a base of support for change. Particular attention must be paid to developing in every elementary school leaders who have specific knowledge of mathematics education. Such leadership is a vital factor in successful change.

Providing Recognition for Accomplishments

In surveys of what employees want most from their jobs, a factor that rates high is full appreciation for work done (Hersey and Blanchard 1977, p. 47). In other words, a little praise for a job well done can go a long way. Mathematics educators should take full advantage of this factor in developing leadership and gathering support for change. Although teachers, supervisors, and other mathematics educators are usually not in a position to give rewards in the form of salaries, promotions, and fringe benefits, they are in a position to give praise for accomplishments. They can reward professional accomplishments with letters of recognition and commendation. These letters are most effective if copies are sent to the person's superior and to the personnel department. The copy to the superior has the added effect of helping the superior to recognize leadership potential of which the superior may not have previously been aware. Furthermore, individuals who receive these letters become more aware of their own leadership potential; the next time they are asked for their support and involvement in change, they are even more willing to participate.

Factors That Influence Educational Decisions

Financial Constraints and Political Considerations

Persons in the field of mathematics tend to think logically and feel that decisions should be made on a logical basis. They are sometimes disillusioned when they make a sound, logical case for a particular change but find that their argument seems to fall on deaf ears. They need to realize that educational decisions are often made on the basis of financial constraints and political considerations. This notion may be difficult to understand and accept by persons such as college mathematics educators who are not directly working in the elementary and secondary schools. However, in higher education, as in any sphere of human activity, logic does not always prevail, and political considerations often influence the decision-making process. Perhaps the dilemma is that the political considerations at the higher education level are quite different from those at the elementary and secondary school levels. Mathematics educators who wish to effect change need to examine the forces influencing educational decisions at the various points of decision. If their logical arguments are not effective, then perhaps the decision makers at the particular point of decision are using a different logic. Mathematics educators will be more successful in influencing educational decisions if they identify the forces that influence the decision makers and recognize that because of these forces the decision makers may have very different priorities. Recognizing this, the mathematics educators greatly increase their effectiveness in influencing educational decisions.

Force Field Analysis

Force field analysis is a technique developed by Kurt Lewin that can be helpful to mathematics educators who wish to influence decisions for change (Hersey and Blanchard 1977, p. 122). Lewin assumes that driving and re-straining forces that influence changes are present in any situation. For a particular change to occur, the sum of the driving forces for that change must be greater than the sum of the restraining forces. In order to influence educational decisions, mathematics educators should examine the driving and restraining forces at the point of decision. Then they can work to maximize the driving forces and minimize the restraining forces. For example, consider the driving and restraining forces on the NCTM's recommendation that basic skills in mathematics be defined to encompass more than computational facility. (In particular, basic skills should encompass at least the ten basic skill areas identified by the National Council of Supervisors of Mathematics [1977]). Driving forces include the influence of the mathematics education profession and the needs of society for citizens competent in the broad range of mathematical skills. Restraining forces include the narrow concentration on computation by some back-to-the-basics forces, ignorance of the importance of the expanded concept of basic mathematical skills, standardized testing that concentrates on low-level skills, and the lack of staff development for teaching the broad range of skills. These driving and restraining forces are illustrated in figure 9.1.

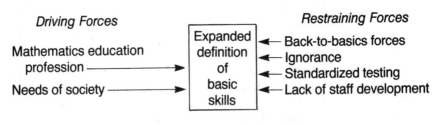

Fig. 9.1

The mathematics education profession can take the leadership in providing information about the needs of society for an expanded definition of basic skills, thereby reducing ignorance and possibly helping to redirect the efforts of the back-to-basics forces. The profession can also influence changes in standardized testing and provide needed staff development. In general, the influence of the mathematics education profession can be the most important driving force for the implementation of all eight of the NCTM's recommendations for school mathematics of the 1980s.

Developing Ownership in Change

Surveys of employees indicate that they have a strong desire to be in on things (Hersey and Blanchard 1977, p. 47). Consequently, they are much more receptive to change when they are able to participate in the decision to make the change. Changes that are sprung without any prior involvement are likely to meet with resistance. People who participate in the planning for change tend to develop ownership in the change, and their support becomes a key factor in successful implementation.

Since curriculum change is ultimately implemented at the classroom level, the support of the classroom teacher is the key to successful implementation. (Center for Instructional Research and Curriculum Evaluation and Committee on Culture and Cognition 1978, p. 19:2). For the change to occur throughout a school's mathematics program, the support of the principal is a critical factor. Therefore, the teachers and the principal should be involved in the planning for change so the change will have their ownership and support. Depending on the nature of the change, involvement and support from such sources as the central administration, the school board, parents, students, or community may be desirable or necessary.

Teachers and administrators must be treated as the professionals they are and not as interchangeable parts of an educational machine that functions according to specifications and procedures that have been determined externally. A school should not be run like a fast-food operation with low-level employees who are expected to perform precisely defined tasks that have been prescribed externally without their involvement. Sometimes elitists outside the schools who think they know what is good for the schools develop programs and attempt to implement them without sufficient involvement of the school personnel. Then, when implementation fails, they tend to blame the staffs of the schools rather than their own ill-considered procedures.

Ownership in change is much easier to develop in a school or school system with a positive climate for change. In such a climate, teachers expect to try new ideas and they find involvement in change professionally rewarding. Their ownership in change and their enthusiasm for change make successful implementation highly probable. Today, teachers tend to be somewhat skeptical of ideas proposed by "experts" outside the schools. However, teachers who are successfully implementing change have credibility with other teachers. The more that teachers have a leadership role in change, the higher the likelihood of success for that change. Teachers tend to react positively to staff development by other teachers and to visits to classrooms where change is being successfully implemented.

A plan for change should include specific procedures for involving those persons whose support is necessary or desirable in the planning process. The time frame for implementation should allow sufficient time for this involvement and for the development of support. It is better to delay implementation to allow time for sufficient involvement than to risk failure with a premature attempt at implementation.

Obtaining Commitment for Change

Getting Started

When the need for change is identified, an assessment should be taken to determine the status quo. This assessment might include data on students' achievement, enrollment, and current practices. Teachers should have the facts and figures about the present situation firmly in hand before embarking on change. Information from an assessment often appears in the form of numbers. This is fortunate because specialists in mathematics have a facility for dealing with numbers. For example, if information from the assessment identifies certain weaknesses in the curriculum or in the allocation of resources, then this information can be used to help establish a commitment for change. Specific goals for change should be clearly established with input from all concerned individuals so that they have ownership in, and commitment to, change. The specific plan for change to rectify the difference between what is and what ought to be should likewise be developed with input from all concerned parties.

Using Force Field Analysis

The leaders who will put the plan into action should do a force field analysis to determine the driving and restraining forces. Then actions should be taken to mobilize the driving forces and deter the restraining forces. For example, a number of years ago, a teacher in Minneapolis wanted to get a computer terminal for his school. He was able to secure the loan of a terminal for several months. He saw that pressure from parents, community, and students was a driving force, and so he proceeded to get his students excited and involved in computer programming. He held open houses for parents, the community, and other students to show them what his students were doing with the terminal. Soon, letters and phone calls started coming in to the superintendent's office. This caused a reordering of priorities, and in spite of restraining financial limitations, funds were found to purchase the terminal. As a result of that experience, it became much easier to obtain funds for computer hardware and service throughout the school district.

Using Information

Data from an assessment can be used as ammunition for change. Sometimes, it is necessary to get data in addition to what is readily available. For example, a number of years ago, the junior high school teachers in Minneapolis complained that their students were coming to them from the elementary schools with inadequate preparation. The city-wide standardized testing program was altered so that students were tested in mathematics at the end of the sixth grade. The test results confirmed the weakness and drew attention. The concern became much stronger when information on each school's performance was published in the local press. As a result, mathematics instruction at the intermediate level started to receive the needed attention and resources, and subsequently achievement improved substantially. Standardized tests have serious limitations for the evaluation of a program because they do not necessarily measure the achievement of the program's objectives. However, the power of the results of standardized tests to influence the administration, the school board, and the community for better or for worse should not be underestimated.

Identifying Decision Points

When conducting a force field analysis, it is helpful to identify the key points of decision for change and to determine which forces are likely to convince the decision makers that change is necessary and desirable. In the situation concerning the computer terminal, the key decision point was the superintendent and the driving force was public pressure. In the example of achievement at the intermediate school level, the key decision points were the elementary school principals and teachers and the driving force was the desire to have their students achieve up to expectations.

A word of caution is in order, however, for leaders who proceed to develop support for change by mobilizing driving forces. A counterreaction can occur if school boards, administrations, or teachers feel that they are being pressured into actions that they do not wish to take.

Dealing with Opponents of Change

Potential opponents of change should be identified, and steps should be taken to minimize the possibility of opposition. This can best be accomplished by involving the potential opponents in the plan for change. When this is done, they often develop ownership in the change and end up providing support.

It is important to avoid getting into win-lose battles over change. Each time a battle is won, increased opposition is mounted from the losers; in the process of winning battles, the war may be lost in the end. If the opposition cannot be persuaded to become involved and it is inadvisable to try to beat them, then what should be done? Probably the best strategy is to go around them. For example, if some teachers or principals are receptive to change and others are not, work with the ones who are receptive. Then, when change has been successfully implemented, the opponents will be in a weakened position. They may either come around to accept and even support the change once its success has been demonstrated, or they may be forced to make the change as the driving forces for change increase. For example, a grades 7–9 junior high school in Minneapolis wanted to diversify its program to meet the varying needs of its students. The school wanted to offer a program in which students could complete the first half of elementary algebra in the ninth grade and the second half during the tenth grade. It also wanted to offer algebra in the eighth grade and geometry in the ninth grade for its more talented students. However, it received opposition from the senior high school, which didn't want to change its program. The junior high school went ahead in spite of the opposition, and by the time it was six months into the program, the senior high school could see the amount of support by parents and students; so it had to go along and modify its program.

If the opposition to a particular change appears to be so strong that the change can be blocked, then the attempt to implement that particular change should be delayed, and concentration should be shifted to other changes. There are plenty of needed changes to go around. As these other changes are successfully implemented, those involved will gain personal power and will then be in a stronger position to go back and effect the change that has been delayed.

Obtaining Resources for Change

Local and External Resources

Often a change requires additional financial and human resources that must be acquired through reordering and reallocating existing resources or securing additional resources. A distinction must be made between the resources needed to make the change and the ongoing resources needed to maintain the change. For example, in implementing the use of computers, the costs of the equipment and the necessary curriculum and staff development for implementation are expenses needed to make the change. Ongoing expenditures to maintain the change include the costs of equipment maintenance and replacement, the purchase of external computer services, telephone costs for time-sharing computer installations, computer supplies such as paper for output devices, and whatever additional human support is needed.

Wherever possible, local resources should be used. This is particularly true of resources that are needed for ongoing maintenance of the change. We can all cite examples of a change falling apart when the extra resources allocated for the change ran out or were terminated. Funds from external sources tend to be available only for limited periods, and so plans must be made to ensure that the change will be maintained after the external funding is no longer available.

Obtaining Local Resources

The situation for obtaining local resources is quite different at the elementary school and secondary school levels. At the elementary school level, the same teachers tend to teach all subjects; therefore, resources for change can be concentrated in mathematics by obtaining a commitment from elementary school principals and teachers that such a change is needed. However, at the secondary school level, even though the mathematics teachers and the principal are committed to the need for change, a concentration of resources in mathematics implies a diminishing of resources in other areas, and that may result in opposition from teachers in those areas. A commitment for change in mathematics as a priority may need to be secured from sources outside the school. Support from the central administration, the school board, or parent and community groups may be needed to obtain the necessary allocation of local resources.

Another strategy for gathering support for resources for mathematics is to structure the change so it will benefit additional portions of the school program. For example, the mathematics supervisor in Saint Paul wanted the high schools to obtain card readers that the mathematics teachers could use with their students to feed programs into the computer. He saw that the schools did not have sufficient money to devote to equipment that would be used exclusively by the mathematics department. Therefore, he developed for the computer a scouting program that could be used by the athletic department, an attendance program that could be used by the administration, and a test-scoring program that could be used by all departments. In view of those multiple uses for the card readers, the schools were willing to allocate the funds necessary for their purchase. In general, as we move toward greater use of computers in mathematics, we should avoid the syndrome of exclusive ownership of computers by the mathematics department. Today, computers have applications in all disciplines for all ability levels of students at virtually all grade levels. If mathematics departments provide leadership in stimulating the broadest use of computers and in sharing computer equipment, they will be able to gather widespread support for the acquisition of computer hardware and services.

Obtaining External Resources

If local resources are not available to effect the desired change, then external funding should be sought. Federal funding can be sought through the Department of Education and the National Science Foundation. Particular attention should be paid to funds that are available through the Department of Education's National Basic Skills Improvement program and Title IV-C program for innovation in education. The National Council of Teachers of Mathematics and the National Council of Supervisors of Mathematics maintain current information about the availability of federal funds. In addition, many school systems have individuals or departments that monitor the availability of federal funds. State departments of education are able to provide information about federal funds and about any special state funds that might be available. In addition, the possibility of securing funds from private foundations should not be overlooked. A word of caution is in order for persons seeking funds for special projects: these funds are usually accompanied by requirements for reporting progress, evaluating results, and disseminating information about the project. These requirements tend to take a lot of the project director's time; therefore, a person who assumes a leadership role in such a project should recognize the priorities and commitment involved.

Local sources such as business and industry should be tapped. For example, in Minneapolis, a local business supplied enough metersticks for all the classrooms and enough metric rulers for all the students. Funds that are raised by parent groups are a potential source for the purchase of computer equipment. In general, local nonschool sources prefer to finance tangible items so that their contributions will be highly visible.

One strategy that can be used in seeking resources is to look for an organization whose goals are very close to one's own. For example, several years ago the Minneapolis schools determined that they needed to increase minority enrollment in high school mathematics courses. At the same time, the University of Minnesota Institute of Technology wanted to increase its minority enrollment. The school system worked closely with the university as they obtained funds from private foundations and industry for an exciting program involving minority students and mathematics teachers from schools in the Twin Cities.

If at First You Don't Succeed . . .

If one is not successful in the first attempt to secure funding for change, one should not give up. Three general restraining forces for change are a lack of money, a lack of imagination, and a lack of determination. However, with enough imagination and determination, the money can usually be found.

Developing Support through Planning

Developing Support through Initial Planning

At the beginning of the planning process, the leadership should identify the needed changes and the driving and restraining forces for those changes.

Priority should be given to those changes for which the driving forces are stronger than the restraining forces. An important driving force for change is a need for change that is seen by teachers, principals, and administrators of the central office. Surveys and meetings can be used to gain information about perceived needs. The information gained can be extremely useful for establishing priorities. When a change is made in response to needs that are identified by faculty, the change begins with broad support and ownership. The change process will be helped if previous efforts have been made to develop leadership for change in a positive climate for change.

The leadership for change should clearly define the goals for change, conduct an assessment of the status quo, and conduct initial planning for change. This information should be presented to the parties who are interested in the change or who will be involved in the change. As the saying goes, "Run it up the pole and see if anyone salutes." If everyone salutes, then go. If suggestions for modification are made, incorporate them if at all possible. In the process, the plan will probably be improved and additional ownership will be gained. If no one salutes, then either put the plan on the shelf or go back to the drawing board. Do not take the attitude that persons who oppose the change are merely negative-thinking obstructionists. Good reasons probably exist for their opposition. Take their concerns into account and weigh them carefully before proceeding with the next step. If their objections are carefully considered and used before the next move is made, the objectors may be turned into owners and supporters of the change.

Developing Support While Refining the Plan

When the proposed change has been reviewed by the interested persons and has received favorable response, the leadership should refine the plan. The leadership that develops and refines the plan should consist either of an individual or a relatively small task force. Members of the task force should be chosen primarily for their expertise in the area of change. However, persons will occasionally be chosen for political reasons because their support will be critical to the success of the change. When the plan is refined, it should be reviewed by the concerned parties once again. Then the plan will be finalized on the basis of their input. In the process, more ownership and support will be gained.

In planning for change, leaders want input from as large a group as possible in order to obtain ideas, ownership, and support. However, a group with representatives from all the concerned constituencies can become very large and unwieldy. Essentially, a large group can do two things in the planning. One is to generate ideas for the plan; the other is to give reactions to drafts of the plan. The actual development of the plan should be done by an individual or a small task force.

It is wise to avoid the dilemma of having to refuse someone's work because it is not of acceptable quality, which might turn the person from a supporter with ownership in the change into an opponent. This dilemma can be avoided by carefully selecting persons to do development work and by establishing a professional atmosphere where criticising work for the purpose of improving it is encouraged and welcomed. A good strategy is to see what people can do with small tasks before assigning them larger ones. A small product can be accepted and then revised, if necessary, without too much difficulty.

When working with a group of teachers on change, leaders will find it helpful to maintain the distinction between curriculum development and staff development. In curriculum development the product is the important thing, and so curriculum developers should be carefully selected and their work accepted according to rigorous standards. In staff development, however, the process is important and the broadest possible participation is desirable. Positive reinforcement should be given for involvement, and criticism that may turn the participants off should be avoided.

Developing Support during Implementation

In planning for change, leaders should give priority to those components of change that have the highest probability for success. For example, initial implementation should occur in situations where teachers and principals are committed to, and prepared for, the change. Then, if the change is not successful, one will not have to wonder whether it was a bad idea or whether it was a good idea that was poorly implemented. If the change is successful, one will have an example of success for everyone to see, and the teachers and principals involved will have increased ownership and commitment. They can then be influential in persuading and training others to make the change.

Uneven implementation can result in increased demands and support for change. For example, suppose a goal has been made of installing five microcomputers in each of five schools, and in the first year funding has been approved for ten microcomputers. The fair thing to do would be to put two in each school. However, the demand for the remaining fifteen might be increased by placing all ten in two of the schools. The resulting creative frustration could be a driving force for further implementation.

Curriculum development should be scheduled with a priority for early completion of some components. This will demonstrate that the project is productive, and the success will tend to develop support for the project; an eagerness will arise to have the remaining components completed. Curriculum development projects are often funded on a year-to-year basis, and funding for a subsequent year is much easier to obtain if completed portions of the project can be demonstrated.

Priorities for implementation should be for those components that require the least amount of change on the part of teachers and that will produce the greatest impact. For example, in Minneapolis, skills-maintenance programs were found to produce substantial increases in student achievement. A skills-maintenance program merely consists of a brief daily review of a variety of previously learned skills. Even the most conservative teacher can and will implement such a program with relatively little training needed. Teachers' support tends to grow when they find that skills maintenance is an excellent transition activity that gets students settled down and working at the beginning of a period of instruction. Therefore, one can demonstrate early success and develop the support of faculty for change by starting with the implementation of a skills-maintenance program or a similar program.

Timing is important in planning for change. The driving and restraining forces sometimes ebb and flow, and, whenever possible, implementation should be done when the conditions are most favorable. For example, positive change is difficult to implement at a time when a school system is undergoing a severe budget crisis. However, that is not the time to give up, either. It is a time when planning for future change can take place. During a budget crisis, many programs must be cut back. However, when the crisis passes, the growth in programs may not be in the same areas that were cut back. With the increasing awareness of the importance of mathematics, well-planned developments in the mathematics program have a good chance for receiving priority attention after a budget crisis has passed.

Avoiding Counterproductive Change

Sometimes the curriculum is driven by forces for counterproductive change. For example, schools are sometimes pressured to adopt educational fads that receive a lot of attention in the media or that gain support through skillful public relations efforts. These fads are often purported to be panaceas that will solve the problems of the schools. School personnel can find themselves on the defensive when asked why they are not implementing these panaceas. One problem is that the wrong questions are asked. Change should occur from defining the problems, looking at various alternatives for solving them, and then selecting the best alternative. Fad pressure starts with a purported solution that the schools are pressured to adopt. The point of attack is usually through the administrative structure rather than through the curriculum structure where the personnel have a knowledge of better approaches. Elementary schools are more susceptible to this pressure because they often do not have sufficient faculty strength in mathematics education to evaluate the fad and provide a response.

Pressures for change, even counterproductive change, must all be taken seriously. School personnel must become knowledgeable about purported panaceas and be able to respond to demands for their implementation. If they display ignorance, they lose credibility. One tactic to ward off the implementation of unwanted curriculum change is delay. A committee to study the change can be set up, and by the time the study is completed, the pressure may have subsided. These pressures do not tend to sustain themselves over long periods of time. Also, given time, the fad may demonstrate its ineffectiveness in other settings, as happened with performance contracting. Another tactic is to work with the persons who are pushing the fad, to demonstrate that a problem-solving approach is being used and that the fad they are promoting may not be the best solution. It may even be possible to redirect this negative force into a positive force for constructive change. If the force for implementation cannot be delayed or redirected, then perhaps the implementation can be limited to a pilot study. At the same time, some pilot studies of better solutions to the problem can be initiated. It will be very rewarding when it can be demonstrated that "brand X" is superior to the purported panacea.

Developing Support through Public Relations

Developing Support through Open Lines of Communication

Open lines of communication are essential for maintaining and building support. People who are kept informed and continually provided the opportunity for input are likely to develop some ownership and to support the change. People who are kept in the dark may get suspicious and tend to oppose the change. With open two-way lines of communication, leaders will be able to identify persons who support change and persons who have some concerns and may become opponents. Then, the leaders can arrange for more involvement by the supporters, thereby building their ownership. Leaders can also modify the plan for change in light of concerns. As the plan is changed, persons who suggested the changes will tend to gain ownership. It may be profitable to involve persons who raise concerns in the planning and implementation, thereby gaining their ownership and support. All persons who express concerns about the change should receive responses from the leadership that their concerns are being considered. Preferably, this response should be in person or by telephone to facilitate give-and-take discussion; then misapprehensions can be identified and dispelled. If the plan cannot be changed in light of the expressed concerns, then an explanation is in order. If logistics prevent a direct personal response, then the written response should be individualized and to the point.

Developing Support through Personal Contact

Leaders for change should practice positive, personal public relations. They should show consideration for others, and they should greet people with a warm smile and welcome. They should make people happy to see them coming. They should make a special effort to learn names, particularly the names of persons whose support is needed for change. Good personal relations can contribute to a person's referent, or personal, leadership power.

Direct personal contact should be used as the means of communication when people are being persuaded to make a commitment to change. This personal contact can come through meetings, individual conferences, or telephone calls. Through personal contact, leaders can respond directly and immediately to any concerns that are raised. Well-executed, direct personal contacts can be extremely effective for building support for change and for increasing personal leadership power.

Developing Support through Written Communication

Written communication can be effective for maintaining support and providing information. Support can wither away if it is not maintained. Written communication is an effective means of maintaining a flow of information about the change to interested parties, thereby maintaining their support. Written communication can also be used to confirm in writing what was agreed on through personal contact, thereby reducing the possibility of misunderstandings.

Brief written statements of fact about the change can be used to build support. For example, a brief statement of the goals of the change and a proposed plan can be used as a handout in meetings where support for change is being solicited or in mailings for purposes of information. Printed materials to support change should be clear and concise. As a rule of thumb, people tend to read what appears on one side of a page immediately. Longer materials tend to be set aside and often never looked at again. It is helpful to have a printed document on hand when one is asked what is being done about the problem. Credibility is gained when one can present written documentation that the problem is being addressed and that there is a plan of action for solving it. As a change progresses, concise progress reports can be used to provide information on progress and to maintain support.

Printed materials that are used to help build and maintain support for change need not be fancy. Simple reproduction of typewritten material is sufficient, although some headlining can be used to highlight important points. Charts can be used to show the timetables for development and implementation. Graphs or tables can convey information on the initial assessment or on progress. The charts, graphs, and tables don't need to be elaborate—just clear.

When a project is completed, printed material may be needed for purposes of disseminating information. Brochures for dissemination should be graphically appealing, and sometimes professional help may be desirable in designing them. Funding agencies of special projects often give suggestions and help for the design of dissemination brochures.

Developing Support through Group Presentations

At various stages in the change process, the leadership may want or need to make presentations before groups to build or maintain support. Such presentations are usually more effective if audiovisual materials are used. Sometimes demonstrations have appeal, particularly demonstrations involving students. Overhead transparencies or slide projections can be used to help with the pacing, clarity, and visual impact of these presentations. "Canned" slide/tape or other media presentations can be used for the maintenance of support or for purposes of disseminating information. However, the personally spoken word of a leader for change is an important ingredient of a presentation to develop support for change. The leader can focus on the specific needs and interests of a particular audience and can personally interact with the audience in a dynamic way to build support. Although presentations that describe programs or projects are sometimes disparagingly referred to as "dog and pony shows," they can be influential in building and maintaining support for change.

Developing Support through Publicity

When change is initiated, publicity about the change should generally be low-key. Constructive change is initiated as a result of a perception that things could be done better. Usually we don't like to advertise that things are not as good as we would like them to be. Exceptions to the idea of low-key publicity at the beginning of change occur when publicity is needed to gather initial support or when the change is being made in response to public pressure; then the schools want to indicate that they are taking constructive steps for change. Educators often get themselves into a trap through premature publicity about the changes they are initiating; then they need to have the evaluation and subsequent publicity support their decision to implement their purported solution to the problem. In the glare of publicity, it becomes more difficult to design evaluation procedures objectively and then use the information to make needed modifications or to discontinue programs that are not effective. In the absence of publicity, it is easier to identify mistakes and quietly bury them.

Once there is evidence that change has been successful, it is important to let the world know. Information about success can help increase credibility and build support. That is why change should be planned so that there is a high probability of some demonstrable success at an early date. Publications of school systems are always ready and eager to publish information about success. If the information is to appear in the public press or the electronic media, it must be considered newsworthy. Many systems have the full- or part-time service of persons who specialize in publicity. They can help with press releases and other means of getting publicity in the press and the electronic media.

Leaders in the change process should develop good working relationships with the press, particularly the education writers. Stories about successful change are likely to receive favorable attention when credibility has been built up. But the press cannot be expected to report only the successes in education. One should also be prepared to deal with the failures and per-ceived failures if the press inquires. When the press is dealt with, it pays to be straightforward. If there are problems, don't try to be evasive or cover them up. Members of the press, like anyone else, are likely to be supportive if problems are frankly admitted and plans for solutions are given. But if it appears that something is being hidden, that one is unaware of problems, or that there are no plans for dealing with the problems, then a difficult time from the press will be deserved.

Finally whenever credit is given for success, be sure to see that the credit is profusely given to all who deserve it. The positive reinforcement they receive will build even greater support for change.

Conclusion

Change is easier in a positive climate for change where new ideas are encouraged and people feel that they can take risks. Leadership for change in mathematics education should concentrate on building expert power, information power, and referent (or personal) power. Educational decisions are often influenced by political considerations and financial constraints. Mathematics educators can gain information for influencing educational decisions by conducting a force field analysis to determine the driving and restraining forces operating at the points of decision. The influence of the mathematics education profession is a key driving force for the implementa-tion of the NCTM's recommendations for school mathematics of the 1980s.

People who develop ownership tend to support change. Commitment for change can be gathered through involvement that helps to develop own-ership. Commitment can be obtained through a strategy of mobilizing the driving forces and neutralizing the restraining forces. Imagination and de-termination may be needed to obtain needed resources for change from local or external sources. Support can be gained through a planning process that assigns priorities to those components of the plan that have the highest probability of success.

Public relations for change should stress open lines of communication and direct personal contact to build support. Initial publicity should be low-key, and then successes should be widely publicized. Persons responsible for successful change should be given positive reinforcement including credit in publicity.

Conditions are favorable for gathering the support necessary for imple-menting the NCTM's recommendations for school mathematics of the

1980s. Two important driving forces are the awareness of the public and the unity within the mathematics education profession. In the 1979 Gallup poll on public education, mathematics topped the list of essential subjects, with 97 percent of the respondents indicating that mathematics is essential (Gallup 1979, p. 40). As we enter the decade of the 1980s, there is widespread agreement within the mathematics education profession with respect to goals and priorities. By marshalling the driving forces of public support and a profession united, we shall be able to gather the support needed to implement the NCTM's recommendations for school mathematics of the 1980s.

REFERENCES

Center for Instructional Research and Curriculum Evaluation (CIRCE) and Committee on Culture and Cognition (CCC), *Case Studies in Science Education*. Champaign-Urbana, Ill.: CIRCE and CCC, January 1978.

Gallup, George H. "The Eleventh Annual Gallup Poll of the Public's Attitudes toward the Public Schools." *Phi Delta Kappan* 61 (September 1979): 33–45.

Hersey, Paul, and Kenneth H. Blanchard. *Management of Organizational Behavior: Utilizing Human Resources*. 3d ed. Englewood Cliffs, N.J.: Prentice-Hall, 1977.

Hersey, Paul, Kenneth H. Blanchard, and Walter E. Natemeyer. *Situational Leadership, Perception and the Impact of Power*. La Jolla, Calif.: Learning Resources Corp., 1979.

National Council of Supervisors of Mathematics. "Position Paper on Basic Skills." *Arithmetic Teacher* 25 (October 1977): 19, 21–22.

National Council of Teachers of Mathematics. *An Agenda for Action: Recommendations for School Mathematics of the 1980s*. Reston, Va.: The Council, 1980.

10

Evaluating Change

JAMES N. RETSON

C ONSCIENTIOUS mathematics educators considering the implementation in their classroom, school, or district of a mathematics program based on the needs identified by the NCTM and its recommendations must consider the process of evaluation as they plan their new programs.

Without a carefully planned, comprehensive design for evaluation, a newly implemented mathematics program could become an academic exercise in futility. Evaluation must become a vital component of changed programs to ensure that (1) the goals and processes are defined and reaffirmed by the users; (2) a program's success is substantiated and documented; (3) the specific procedures followed in the implementation of the program are documented; (4) valid data are available if comparisons to other programs or past performance of students are warranted; and (5) changes, corrections, and improvements can be made as needed. In addition, data and results from an evaluation are necessary if the improved program is going to be adopted by others. Changed programs can be proved and better explained with usable information from an evaluation. Conversely, a program thought to be successful by the local implementers will probably remain local or be quietly discontinued without results from an evaluation.

This chapter considers the essential components of an evaluation of change and presents a suggested model for evaluating the effectiveness of a changed program in mathematics.

It is suggested that an individual or team (depending on the magnitude of involvement) be specifically assigned to the task of evaluation and management. This person's or team's responsibilities would include the following:

1. Assisting the staff in identifying their particular needs and reaffirming their goals

2. Meeting at regular intervals to review progress and recommend needed adjustments in the program

3. Assisting the staff with the selection or development of appropriate tests, checklists, and other instruments for evaluation
4. Observing and monitoring the program and documenting those observations
5. Expediting the necessary testing of the students in the program
6. Collecting, analyzing, and sharing all results with the staff
7. Reporting these results to the appropriate agency or authority

This designated individual or team can be most effective to a program, since the evaluative function can occur with minimal time expended by teachers and students, who can then concentrate on the teaching-learning process.

Evaluation Inquiry

Before developing a strategy of evaluation for measuring the effect of change, the staff who are planning a change in their mathematics program must first ask some key questions related to the new program:

1. How is our new program different from what we have been doing?
2. After we have implemented the new program, how will we know if it is successful?
3. What did we do in our curricular practices to achieve success in the new program?
4. Was our program well managed and did we accomplish what we set out to do according to a time schedule?
5. How did our students feel about the new directions taken?
6. How did the teachers and aides feel about the new program?
7. Did we communicate the new program to parents, students, community, other teachers, and administrators?

Seeking answers to the sample questions above *is* evaluation and represents a fairly broad range of concerns about evaluation. The evaluation inquiry will, of course, vary depending on the particular school situation.

While seeking answers to these questions, the evaluator/manager, together with the staff, must employ two basic *types* of evaluation—*process* and *outcome*—which, when done properly and combined, will give all the information needed to evaluate the new program.

Process Evaluation

If we consider evaluation in terms of the "means" and "ends" dichotomy, process evaluation is the means and outcome evaluation becomes the ends.

Concerns about process in an educational program would include such items as program-management tasks and time constraints for completing these tasks; instructional concerns including teaching styles used, instructional materials, time spent on mathematics instruction, the classroom environment, instructional methodology, and the nature of students in the classroom; and in-progress revisions of activities and products developed.

The seven *process* concerns listed above more clearly define process as those elements within a program that contribute to the results or *outcomes* reached in the program. To put it another way, process is defined by Tuckman (1979, p. 299) as "how the program operates in the classroom or school; that is, the classroom behavior of teachers and students involved in the program." The process evaluation should provide a rich description of the new program that will assist others who seek to replicate or adopt it.

Measuring Process

Process evaluation of a program is usually done through some form of monitoring or record keeping of the program's activities. These activities are either management-related activities or instructional activities. A process evaluation of management-related activities can be documented using a form similar to the one in figure 10.1 (filled in with hypothetical tasks).

The completion of a management monitoring form serves some important purposes for a project and its staff, since it confirms and documents that all management and evaluation tasks have been completed on time; it keeps all involved individuals aware of task-time expectations; it provides continuous information regarding both progress and needed revisions and other adjustments; and finally, it provides management information that can ensure more efficient replication in a new setting.

A process evaluation of *instructional* activities is best accomplished through systematic monitoring. This monitoring task should be done by the assigned evaluator, other trained persons, or the project manager and can be accomplished in a variety of ways and at different monitoring intervals. When the instructional concerns mentioned above are used, a checklist instrument such as the one in figure 10.2 could serve as another instrument for the systematic compilation of process data.

The instruction-related observation form together with the management-related form can yield most of the data needed for the process evaluation component. These data can be compiled at the end of a longer period of time (e.g., full year of implementation) when outcome information from students is available. An analysis of a program's processes and related outcomes will give the staff many systematically determined insights into their program.

These observation forms, filled out at least four times a year in each project classroom or a random sample of classrooms, help the staff deter-

Management Monitoring Form

(Date)		(Project Evaluator)		
Activity	Person(s) Responsible	Date of Completion	Present Status	Comments and Suggestions
1.0 Hold meeting to define goals of new program.	All staff and evaluator/ manager	8/30/81	Yes	Two staff members not in attendance. Hold individual sessions.
2.0 Hold periodic meetings with staff to discuss progress and revisions.	All staff and evaluator/ manager	10/15/81 12/15/81 2/15/82 4/15/82		
3.0 Select or develop tests, observation forms, sur- veys, and checklists.	Evaluator	8/2/81		
4.0 Administer pretests and posttests of skills.	Teachers	10/1/81 5/15/82		

Fig. 10.1

mine if change is indeed being implemented, how it is being implemented, what changes need to be made, and to what the outcomes can be attributed.

Outcome Evaluation

Outcome evaluation determines if the goals and objectives of the new program have been attained by students; it constitutes the "bottom line" purpose of any educational program. Students' achievement cannot be measured, however, unless the objectives to be achieved by those students are specified at the time a new, changed mathematics program is in the planning stages. These objectives, when placed in a somewhat sequential order, are sometimes referred to as a continuum of competencies to be attained. Whatever they are called, outcomes cannot be measured without these prestated objectives. Many school districts have already formulated their own continua of objectives that can be used for purposes of evaluation. It is important for the staff to make certain that existing continua contain the new goals and objectives inherent in the changed program. Staffs whose new programs draw on the NCTM's recommendations might include in their continua objectives related to problem solving, basic skills beyond mere computational facility, and computer literacy, to name a few.

After these competencies have been defined, added to, or formulated anew in a changed mathematics curriculum, the formulation of a test based on these competencies is in order. The purpose of this test is to measure long-term achievement of students at the end of a full year or semester of implementation. This test should be developed locally to ensure that test items directly reflect the objective of a particular program. The test will be known as a criterion-referenced test, since the test items *refer* to objectives or *criteria* of the program.

Many commercially prepared and published standardized (norm-referenced) tests are available for use in determining long-term achievement. These are satisfactory if the users are convinced that the test items are a true representation of their own program's specific objectives. Most available standardized tests do not reflect the new goals of a mathematics program. Standardized or norm-referenced test scores might be useful if comparisons to a national norm for an age or grade are made. This information is not necessary for the initial evaluation of a new, changed mathematics program, however.

It is therefore recommended that a criterion-referenced test be developed and used in a new mathematics program. This test will yield information regarding students' achievement to the staff so that they, in turn, can make decisions about the success of their new mathematics program. The results of these outcome measures are usually expressed as a percentage of items achieved by each student and can be analyzed in terms of individual scores or a mean (average) score for a group.

An expected score should be determined by the staff prior to the implementation of the program. The agreed-on expectancy will be a score that, if attained, will constitute the success of a program and should be stated as an objective of the new program.

Examples of expected objectives might include these:

1.0 After one full year of involvement in the new mathematics program, all eighth-grade students will demonstrate achievement by scoring 85 percent or better on a criterion-referenced test designed to measure the achievement of mathematics objectives at their grade range.

or

Monitor
Process Instrument (Mathematics)

School _____   Date _____

No. students in class _____   Teacher _____   Grade _____

Students' abilities (high, average, low)

1.0 **Materials of instruction**
Check if available and in evidence
  1.1 Calculators   _____
  1.2 Computers   _____
  1.3 Textbooks   _____
  1.4 Books other than texts for instruction   _____
  1.5 Manipulatives   _____
  1.6 Games   _____
  1.7 Other _____

  _____

  _____

  Comments _____

  _____

2.0 **Instruction**
Indicate evidence of the following:
  2.1 Individual needs of students are being accommodated (e.g., different pacing allowed, small-group or individual instruction, independent study, etc.). Yes _____ No _____
  2.2 Average number of minutes per day spent in class on mathematics _____
  Outside of class (estimated) _____
  2.3 Objectives in mathematics for students are available.
  Yes _____ No _____
  2.4 These objectives are known by the students.
  Yes _____ No _____
  2.5 Basic skills considered include (check those in evidence)
    2.5.1 Problem solving   _____
    2.5.2 Estimation   _____
    2.5.3 Measurement   _____

    2.5.4 Computer literacy   _____
    2.5.5 Predicting the reasonableness
           of results   _____
    2.5.6 Geometry   _____
    2.5.7 Computational skills   _____
    2.5.8 Use of tables, charts, and graphs   _____
    2.5.9   _____
    2.5.10   _____
  2.6 Teaching strategies include
    2.6.1 Lecture   _____

2.6.2 Demonstration   _____

2.6.3 Questioning   _____

2.6.4 Independent projects   _____

2.6.5 Reading   _____

2.6.6 Field experiences   _____

2.6.7 Other _____

Comments _____

_____

3.0 Classroom environment

3.1 The classroom is generally comfortable and not insti-tutionlike.   _____

3.2 Games and manipulatives for mathematics are evident.   _____

3.3 Books and other materials are readily available.   _____

3.4 Bulletin boards are interesting and are related to mathematics.   _____

3.5 Students' work in mathematics is displayed in the room.   _____

3.6 Areas are available for small-group work.   _____

4.0 Parent involvement

4.1 A means by which students' progress is communicated to par-ents   _____

4.2 A means by which the objectives of the program are com-municated to parents   _____

5.0 Staff development

5.1 The teacher feels competent and comfortable with the program. Yes _____ No _____

5.2 The teacher feels the program could be improved in the following way(s):

_____

_____

_____

5.3 The teacher would like more information or expertise in the following areas:

_____

_____

_____

Additional comments: _____

_____

_____

6.0 The attitude of the teacher toward the program can be considered (check one):

☐ very favorable   ☐ good   ☐ fair   ☐ apathetic   ☐ negative

Fig. 10.2

2.0 After one full year of involvement in the new mathematics program, eighth-grade students will have achieved a *mean* score of 85 percent or better on the criterion-referenced test.

or

3.0 After one full year of involvement in the new mathematics program, at least 90 percent of the eighth graders will achieve a score of 85 percent on the criterion-referenced test.

After the results of these outcome measures are made available to the staff by the evaluator, it is left to them to review the results and either rejoice or rationalize!

Attitudes

A new or changed program should consider the attitudes of students and teachers. If staffs are interested in upgrading their mathematics program, either by enriching the content or by revising the process, they should seriously concern themselves with students' attitudes toward that new program. Mathematics phobia or anxiety may also need to be considered.

The attitudes of students can be a determining factor in how successfully the objectives of a program are achieved. By the same token, good achievement by students can be a determiner of attitude. Despite this "chicken or egg" argument, attitudes as well as achievement should and can be measured. We would hope that a new program is motivating and will "turn on" students.

Measuring attitudes need not be a highly sophisticated process; attitudes can usually be determined by a simple homemade questionnaire administered to students as a preprogram and postprogram measure. An example of an instrument to measure changes in attitude toward mathematics is given in figure 10.3.

This sample questionnaire on attitudes would be scored accordingly to how often mathematics appears in the responses. For questions 1, 3, and 4, give five, three, and one point if mathematics appears first, second, or third, respectively. Subtract five points if mathematics is inserted in question 2 and give five points if it is inserted in question 5. The item scores are totaled for both the pretests and posttests, and an average score is computed for a group (class) for each test.

We would, of course, anticipate a higher average score on the posttest than on the preprogram measure. If this occurs, we can consider that a pattern of improved attitudes toward mathematics has been achieved. The degree of improvement from pretest to posttest is relatively unimportant, since we are not interested in statistical analysis of results, only that a positive trend has been established. The instrument in figure 10.3, we

---

Questionnaire on School Subjects

Name _____ School _____ Grade _____
        (optional)

Note: This is not a test and will not be graded. There are no right or
       wrong answers. We are only interested in *how you feel* about
       school subjects.

Please answer the following questions.

1. List your first, second, and third most favored subjects in school.

    1. _____
    2. _____
    3. _____

2. If you could eliminate one subject from your school program, which
   would it be?

   _____

3. Which subjects (first, second, and third) will best prepare you for a
   job?

    1. _____
    2. _____
    3. _____

4. Which subjects are the most help to you in your daily living now?

    1. _____
    2. _____
    3. _____

5. In which subject do you feel you learn the most?

   _____

Fig. 10.3

reiterate, is a simple example of what *might* be used to give some indication
of how students feel about mathematics in their school program.

Another method of determining a change in attitudes is through the use of
a *semantic differential*. An example of an instrument of this type is shown in
figure 10.4. This instrument should also be administered to students, by
classes, as a preprogram and postprogram measure (September and May).

Teacher _____ Grade _____ Date _____ Student _____

_____ (optional)

For each of the six items below, check the *one* term that best describes how you feel about mathematics.

VALUABLE	very valuable	valuable	sort of valuable	neither valuable nor worthless	sort of worthless	worthless	very worthless	WORTHLESS
UNPLEASANT	very unpleasant	unpleasant	sort of unpleasant	not pleasant or unpleasant	sort of pleasant	pleasant	very pleasant	PLEASANT
INTERESTING	very interesting	interesting	sort of interesting	not interesting or boring	sort of boring	boring	very boring	BORING
DIFFICULT	very difficult	difficult	sort of difficult	not difficult or easy	sort of easy	easy	very easy	EASY
LIKE	strongly like	like	sort of like	neither like nor dislike	sort of dislike	dislike	strongly dislike	DISLIKE
UNNECESSARY	very unnecessary	unnecessary	sort of unnecessary	not necessary or unnecessary	sort of necessary	necessary	very necessary	NECESSARY

Fig. 10.4. A semantic differential to determine change in attitude

This particular semantic differential contains seven choices for each item and can be scored accordingly: 7 points for the highest response and 1 point for the lowest. An average score for a class can then be determined, and one would assume that scores on a postmeasure would surpass the premeasure score. Note that on this example every other item is reversed, and scoring should be done accordingly.

Teachers' attitudes toward the program can be determined from information gathered on item 6.0 of the monitor/process instrument form in figure 10.2.

Measuring Communication

One of the questions included under "Evaluation Inquiry" considered the matter of communication. The success of any new program and the establishment of that program will be better ensured if the lines of communication are kept open.

All persons who might be affected by a new or changed program should be kept informed or become indirectly involved in that program. These include teachers implementing the program, other teachers indirectly involved, administration, parents, interested members of the community, local media, and, of course, students. Evaluating the communicative aspect of a program presupposes that in the planning stages of that program, provision was made for communication activities (e.g., regular meetings with involved staff, meetings with parents to explain the program early in the school year and perhaps again at midyear, press releases as deemed feasible and timely, meetings with other teaching staff and administrators, and, finally, keeping students informed through the classroom teachers' explanations of goals and expectations).

Evaluating these efforts at communication is a simple procedure of documentation. Documentation consists of some form of *written* record or log of meetings, contacts, and presentations. This record should include dates, the nature of the meeting or event, persons attending, and outcomes, if appropriate.

It is, of course, recommended that one individual (manager/evaluator) assume the responsibility for keeping and compiling the log of communication events. This process of keeping logs and records need not be a cumbersome task and should be simplified as much as possible. The act of compiling and documenting this information will not only serve as a reminder to staff that communication events are to take place but also ensure that these additional process data will be available for use when determining how effective the new program has been and what decisions should be made regarding the further implementation of change.

Using the Data

After process, outcome, and other data have been collected as suggested, many tentative conclusions and future decisions regarding a program can be made. The strategy employed for evaluation will, of course, vary depending on the unique needs of each new program. Although the collection of process and outcome information suggested in this chapter is not considered highly technical in the sense that statistical analysis is appropriate (i.e., no reference is made to validation of instruments, extensive use of norm-referenced tests, comparison groups, and statistical determinations), the information is legitimate for systematically determining if the initial evaluative questions have been answered.

Evaluating change in the manner suggested in this chapter may seem rather cumbersome and require more paperwork than some would deem necessary. This is not necessarily true if good planning is done, and the task will be minimal if one person or a small team is assigned as coordinator(s) of the evaluation effort. Completing a systematic evaluation of the change will make legitimate the effort for change, and the entire venture will be a more enjoyable and satisfying experience for all concerned.

Summary

The strategy used for the evaluation of change is quite different from research-oriented evaluation. The evaluation of a unique educational practice or program is usually thought of as a measurement of outcomes and a statistical comparison of those outcomes to those of another, nontreatment group. In evaluating change, we are, of course, interested in outcomes but only to the extent that those outcomes are equal to, or better than, predetermined expectations of that changed program.

Effective evaluation of change will require, in addition to measures of outcome, a consideration of process and formative evaluation. The data and the analysis of that collected data will seem rather informal, but they are legitimate and do not necessarily require the services of persons highly experienced in evaluation.

A sequence of events suggested for evaluating change should include the following:

1. The formulation of questions to be answered through evaluation
2. A clear understanding of the discrepancies identified from needs and recommendations
3. The listing of specific changes proposed for the program
4. The scheduling of regular meetings for the purpose of reviewing the goals of the change

5. The development of a specific continuum of objectives for students that incorporate those mathematics competencies represented in the new program and statements of those expectancies

6. The construction of a criterion-referenced test that is based on those objectives

7. The development of *process* forms to be periodically completed by the evaluator

8. The development of a survey or questionnaire of attitudes and its administration to students as a preprogram and postprogram measure

9. Planning, scheduling, and recording the efforts at communication included in the program

10. Compiling and comparing all data collected

11. Using all information collected and analyzed to make decisions about the continuation of the program

Finally, it is strongly and seriously recommended that the entire involved staff celebrate their success at the end of the first year of the new program!

REFERENCE

Tuckman, Bruce Wayne. *Evaluating Instructional Programs*. Boston: Allyn & Bacon, 1979.

PART

Changing, and Being Changed by, Others

Grant me the serenity to accept the things I cannot change,
The courage to change the things I can,
And the wisdom to know the difference.

Espoused by many groups and found in many forms, the quotation above should be the guiding thought for any person contemplating changing mathematics programs. In Part Three, Osborne discusses those outside areas in which educators have little effect but that themselves may bring about changes in the schools. He also points out that educators must be alert to taking advantage of these external forces. Price considers those areas in which educators can have some effect and makes recommendations for actions that can be taken. Sobel summarizes Part Three by pointing out how teachers can implement the *Agenda for Action*, citing specific resolutions and actions. His positive approach to change and to the role of the teacher in it concludes the book serenely, courageously, and thoughtfully.

161

11

Change and the External Environment

ALAN R. OSBORNE

M ATHEMATICS educators find it easy to consider educational change in terms of their primary interests—mathematics teaching and learning. However, the educational enterprise and the forces operating on it encompass more than mathematics education. Many changes in schools and schooling take place outside the immediate interests and locus of control of mathematics educators. Many of these external changes and the forces and situations that spawned them have had a profound impact on mathematics education by producing or causing reactive changes in mathematics programs. In order to understand some types of change in mathematics education, it is necessary to appreciate the environment external to mathematics education and the major categories of change that operate within that environment.

Five major categories of forces for change, or sources of influence, in the external environment are considered in the sections that follow:

1. Government agencies concerned with education
2. The law—court and legislative mandates
3. Attitudinal shifts in the community environment
4. Cumulative inattention
5. Private institutions, agencies, and businesses

These categories and their effects on change in the local schools are not mutually exclusive; examples in one category might also appropriately be considered in a different category. The assignment to categories is on the basis of the most salient characteristics of each example for its impact on mathematics education. Each source of influence has a profound effect on the teaching and learning of mathematics, and recent evidence indicates that

163

the effects will require reactions, consideration, and use by mathematics educators in the immediate future.

Government Agencies Concerned with Education

Mathematics educators and other professionals in the schools find it significantly easier to be optimistic about change if they can make positive, forward-looking decisions. Government agencies make available each year a variety of resources for use by the schools. Because of the tradition of local control, these resources must be applied for or requested. The act of deciding to make application for the resources and the identification of the purpose for their use gives the local school system a sense of ownership that can help produce desired change. In addition, the resources thus obtained are an effective supplement and extension of the local resources that can be brought to bear on educational problems. These resources and the processes of applying for them are called *opportunities for change* in the discussion that follows.

It is naive to assume that many opportunities for change are generated by government agencies solely to provide resources for the functioning of local schools. Resources are seldom invested without some purpose. A primary responsibility of local leaders is to assure a match between the purposes of the agency or institution in providing the opportunity for change and the needs, goals, processes, and personnel at the local level. The leaders of the local system must not only be aware of potential sources of aid for the process of change but also have enough information about their local situation to be rational and systematic in judging the appropriateness of the source. Wilkes, Strong, and Coblentz (1977) describe an ongoing needs-assessment process that can be used to ascertain how the context of the local school fits the conditions necessary in requesting resources.

A change generated by outside sources has the potential of making a considerable difference in how well local schools and teachers can function in facilitating children's learning of mathematics and in improving teachers' attitudes and self-concepts. The teacher who participates in a program or uses resources that stem from an outside source usually recognizes that something special and extra is available for use in the education of students. Thus, leaders in local systems will frequently accept an opportunity for a change that does not fit the needs of the school system precisely for the advantage that accrues with the attitudinal shifts of the teachers.

Change that depends on special outside resources is a problem for mathematics educators if the effects are to be lasting. Outside resources are extra and supplemental. If the change is to be permanent, then the planners must recognize what will happen if the supplemental resource is withdrawn. The plan must either produce change in behavior or tools for instruction lasting beyond the period of the special project or include the means of collecting evaluative information to convince individuals responsible for local resources to support the special program with those resources. Thus, three types of outside resource programs deserve high priority:

1. Those that provide the resources to change the behavior of teachers during the course of the program. The outside resources provide the seed money to try what is new and different and highly likely to produce productive, permanent change.

2. Those programs giving outside resources that allow the school system to test alternatives. An important characteristic of this impetus for trying alternatives is the systematic collection of evaluative information that allows comparative decision making about the alternatives and their impact on teaching and learning mathematics. In short, is the change worth it?

3. Those programs that provide tangible equipment and materials.

The goal of most uses of outside resources is to produce a change of some permanence in response to a local need. The orientation for innovation derives largely from the intent of agencies and local school systems to maintain the tradition of local control of education. During the decades of the 1960s and the 1970s, a variety of approaches were directed toward creating educational change and innovation at the local school level. Some principles for using outside resources to produce such change are apparent in this recent history.

The first massive attempt at effecting a change in mathematics and science education through the use of federal resources was the post-Sputnik endeavor of the National Science Foundation (NSF). The primary efforts toward change were through curriculum development in such projects as the School Mathematics Study Group and through in-service education directed primarily at improving teachers' knowledge of the subject matter. Although many postmortems, such as the NACOME report of the Conference Board of the Mathematical Sciences (1975), bemoaned the lack of continuing impact from the NSF effort, the effort established a new role for the federal agencies in attempting to change what was happening in the schools. The effort also revealed some characteristics of the process of change.

Perhaps the most significant lesson of the first federal intervention in the process of educational change derives from the fact that the process did not honor local situations. Teachers who attended the in-service institutes represented many local systems, and the effects of their new knowledge and insights were dissipated. There was no assurance that a school system would have a sufficient cadre of knowledgeable and concerned teachers to carry through or implement an element of change. No organized attempt was made to relate suggested changes to local curricular needs, to the expertise of local teachers, to local curricular plans, or to the perception of need held by the local school community. Most institutes did not address the task of helping participants acquire the leadership skills and understandings necessary to produce a desired change in their local schools. School systems typically stumbled toward a change without being more than superficially convinced of the wisdom of the endeavor and with little expectation of the problems that might be encountered. In short, good intentions were not enough to cope with the difficulties that were encountered or even to identify the difficulties. The evidence indicates that change was envisioned and planned for within the small community of mathematicians concerned with mathematics education (small and local in terms of the communication patterns) without a realization that a change in the schools involves myriad other small, local communities of communication. Each of the local school communities that is the site of intended change must be dramatically involved in the process of change if it is to be successful. The marvel is that as much change occurred as did.

A second retrospective conclusion about the large-scale reform effort of the 1960s was that the designers of the curricular change effort were not adequately systematic and complete in their planning. The curricular changes envisioned for mathematics demanded teaching skills and insights different from those currently in use by the large majority of the teachers. Neither the new teaching materials nor the institute programs sufficiently considered this facet of the implementation of change in their planning. The case might be made that given the state of knowledge about the teaching-learning processes for mathematics and science evident in the 1950s, the process of change concerning changing teachers' behavior was as systematic and rational as could be expected. However, insufficient attention was given to the details of how teachers should change their behavior in the classroom or their thinking about curricular problems. The lesson for current efforts at curricular change is that needs assessments must be thorough with a relatively complete analysis and matching of current understandings and skills of teachers with the demands of those required by the change. A second lesson learned relative to teaching behaviors and the process of change is that the degree of change required and the relative stability of teaching behaviors are quite easily underestimated. That is, an "unfreezing" of teaching behaviors and attitudes may be required.

The second major effort of a federal agency toward implementing educational change that provides insights into the process of change in the schools is the Elementary and Secondary Education Act (ESEA) of 1965. The political pressures that incorporated efforts toward local control and that featured an evaluative component are documented by McLaughlin (1975, 1976). The well-intended features, along with dollars provided by an outside source, represented an ambiguous mixture that has seldom resulted in lasting change. Local control often suffered from the inclusion of uninformed but well-intentioned people in the planning processes. Designing a funding program that had other than a local source but that depended primarily on local evaluation often resulted in evaluative reports capitalizing more on features of projects that would be well perceived by the funding source than on features useful to the local school system in monitoring the process of change or to other school systems in designing and adapting the program to fit their needs. The lesson to be learned from the ESEA experience is that local control can readily be compromised by programmatic features that ambiguously mix the responsibilities and power of the funding source and the local school. Yet it is unrealistic to expect the agency that is making the investment to relinquish the stewardship of the investment.

The opportunities offered by outside agencies for the investment of resources in changing of school mathematics programs are an important feature of planning for change. The resources make it possible for many school systems to try distinctive, innovative thrusts in their mathematics programs. For the majority of school systems this is the sole source of capital for risk or venture in curriculum, instruction, and evaluation. A school system must realize that use of this capital often demands an investment of local resources of an intangible nature, namely, the creative and emotional commitment of local staff. Thus, the school system has a major responsibility to assess whether the goals of the funding agency and the conditions necessary for the application of the funds are consistent with the purposes and program of the local school.

The difficulty in taking advantage of resources available through outside agencies is that many of the high-priority problems and needs of the local school system do not conform to the priorities of the agencies. The decision of whether to apply in the entrepreneurial sense or to wait until the priorities of the agencies fit the school system's needs and problems is both frustrating and difficult. Making progress on a low-priority problem while accruing benefits from the additional resources obtained from a funding source is at least a step in the right direction.

Government agencies provide a resource for facilitating educational change that cannot be ignored. However, the local school system must incorporate its needs and resources into the process of application and must use the knowledge and attitudes of local personnel if the change is to enjoy permanence. Furthermore, information must be collected concerning the effectiveness of the change in terms of the investment in recognition of the possible cessation of the outside resource.

The Law—Court and Legislative Mandates

Imposed change resulting from an action by a court or a legislative body is a factor that mathematics educators must consider in the process of change.

Because a school does not have the option of deciding whether to participate, this form of change provides a particular set of problems for leadership personnel. Teachers and other school personnel usually take more pride in working on a change in response to problems or goals they have helped identify than on a change imposed from the outside. If, for example, a court threatens punitive action as a result of apparent segregation or funds are withheld because of failure to comply with legislation on mainstreaming, the school's reactions are seldom as positive as those resulting from a school's self-determined beliefs and values.

A second significant factor in imposed change is that the change is often highly disruptive of policies, behaviors, and traditions in the local school setting. The mandate for change stemming from court action seldom contains the resources necessary to accommodate or support the change. Legislative action sometimes provides extra resources to support the change but frequently does not. Thus, the school is frequently caught in a trade-off situation in which money or personnel that have served to make the established program work must be shifted to support the change. This contributes significantly to the perception of disruption by the people affected by the change.

The attitudinal factors identified above can dramatically change the atmosphere of a school. Teachers and other affected individuals can view the imposed change negatively to the extent that teaching and learning take place in an unhappy, nonconstructive environment. Energies that should be expended in teaching and planning are dissipated in complaining and protesting. Resisting imposed change typically is a waste of energy representing only a delaying action for the inevitable. Thus, the major tasks of leaders in coping with imposed change are defusing the negative reactions and getting on with the business at hand, namely, providing a constructive learning environment.

There are two keys to producing a constructive environment in reacting to imposed change: (1) sharing information and (2) involving the concerned individuals in the decision making about policies affected by the change. In any process of change, informed judgments based on sound knowledge lead to more productive reactions than those based on hearsay and incomplete evidence. Thus, a leader must be sure to obtain accurate information to communicate. An in-service program may be necessary to allow the participant in the change process to assess more accurately, on both a personal and a system-wide basis, the needs that will contribute to making and understanding those difficult trade-off decisions that may be required. That is, the focus of dialogue concerning imposed change should be on actions to take rather than on regrets or resistance.

If a court or legislature mandates a change, it typically is in an arena of social concern or is symptomatic of a problem of sufficient significance that it deserves attention and thought by school personnel apart from the solution imposed by the court or legislature. Indeed, one may regard many of the rulings and laws as definitions of problems about which educators need to be concerned. Many of the problems of our society, such as the treatment of minorities, equality of opportunity for women, population growth, and mathematical literacy, have resulted in court rulings or legislative actions that tell or reaffirm for educators that schools play an important, significant role in helping shape society and determining an individual's contribution to that society. Although we may not agree with the form that an imposed change takes, it is an important reminder of the stewardship and responsibility given to the schools. Thus, for example, legislation for minimal competency in mathematics is an affirmation of a societal need for a functionally literate citizenry and a testimony to our importance and responsibility as mathematics educators.

If we accept the principle that change imposed by the courts or legislative action is symptomatic evidence of broader concerns and problems in society, then it is incumbent on leaders to assess what significant and important concerns and problems are emerging. This will allow the schools to take anticipatory action in dealing with the societal concerns and problems rather than await the imposition of required change that may take a form uncomfortable or disruptive to the operation of the school. Thus, mathematics educators concerned with educational change must look outside their own field in order to have the judgment necessary to control their own destiny.

A recognition of the importance of an acceptance of the change by the affected or responsible individuals has been featured as a significant factor that shapes the results of the change. This is not to say that a school system or an individual teacher must support the mandated change when it may be apparent that the change may adversely affect the learning of mathematics. Acceptance must be considered in two senses, the pragmatic sense of providing for instruction as opposed to the political sense of trying to revoke or suspend the imposed change. We need not be passive and accepting of a legislative action or a court ruling affecting the performance of our responsibilities in mathematics education. The decision to support or not to support the change and the resultant political actions to be taken are independent of the pragmatic functioning within the change to provide students an opportunity to learn mathematics. Children do not have the opportunity or luxury of considering noncompliance with rulings or new laws; school attendance is required. Consequently, the teacher and the school system must behave in a way that does not compromise the children's opportunity to learn.

Students have no choice and often little power to be other than passive respondents to an imposed change. Assessing the risk of noncompliance necessarily includes the judgment of the effects of noncompliance on the students.

Finally, the failure to offer specific recommendations about how to deal with imposed change is the result of the remarkable diversity and range of concerns about the schools held by society and the many different forms that the mandates for change have recently assumed. Beyond the simple dicta to be informed and rational, it is difficult to specify appropriate recommendations that would fit all situations that will develop in the future. Certainly the trade-off decisions are more difficult because the imposed changes frequently concern fundamental value systems and habituated behaviors that are traditional and unquestioned by many individuals. The effective leader takes advantage of the imposed change insofar as possible to design opportunities for peers and staff to grow and develop in serving the needs of the students and the school. The leader will be realistic and practical in attempting to minimize the disruptive potential of the required change.

Attitudinal Shifts in the Community Environment

The community of a local school can be described in terms of the values held by citizens, socioeconomic status, types of local industry, attitudes, educational level, and myriad other characteristics that can contribute to the process of change. In simplistic terms, shifts in the community environment frequently represent powerful forces for change in the mathematics program. Capitalizing on the community environment to contribute to the process of change relies on an accurate judgment of when a shift takes place and a valid assessment of the nature of that shift. The categories of change discussed previously can be characterized by a decision being made at a specific, identifiable time that requires an action or a reaction by the responsible personnel of the local school. The shifts in the community environment are rarely other than gradual evolutions.

The 1960s began with extensive interest in serving the needs of talented youngsters in mathematics. During the later part of the decade, interest and attitudes indicated greater commitment to the less able and disadvantaged students. The decade of the 1970s closed with a reawakened interest in the talented. The school system that desired to capitalize on the shifting interests of the community concerning special categories of students would have needed to judge accurately when the community interest was sufficient to support programmatic changes. A premature call for action and support appears to the community to be wishful thinking on the part of school personnel, but a failure to promote change until well after community interest peaks contributes to the local community's perceiving the school as less than an exciting and stimulating situation for children and youth. By focusing efforts for change in terms of the peaking interest of the community, a local school can significantly enhance the stature of the school in the community and thereby help to assure a better level of support for the ongoing program.

A fundamental difficulty with using shifts in community attitudes and values to support educational change is that it can result in a faddish, transitory curricular program through the years. To have established a program for the talented student in mathematics in the early sixties and then to have relinquished this program ten years later to serve the needs of the disadvantaged would have been regrettable. Thus, another fundamental responsibility for the local school staff when capitalizing on shifts of community attitude is planning to assure for the continuation of a program when interest wanes. For many of the attitudes and values about schooling, interest sufficient for the support of change appears to last only a limited period of time, two to three years at most.

The shifts in the community environment often have more permanence than is evident in the example given above of serving the needs of special categories of students. Currently the issues and problems of how to best incorporate and use the calculator or the computer are significant and important for most schools. The expected curricular change assumes a greater significance and importance if it is noted that the calculator and the computer are a part of the environment. Most households have at least one calculator. Most adults have at least an indirect encounter with a computer in the course of a week, and an increasing portion of the encounters is the direct interaction between the individual and the computer. The present evidence is that both the calculator and the computer will become even more pervasive in our day-to-day life. In a dramatic sense, the state of the community environment has changed. Unlike the shifts in attitudes and values concerning the design of curricula for special populations, the shift in the state of the community environment relative to calculators and computers can never return to the previous state; it is permanent and lasting.

Few changes in the community environment, however, are of a lasting and permanent nature. Shifts in attitudes and values seldom are. Major new technological tools that have widespread use across the community, such as television or the calculator, usually yield a shift of some permanence. For a small, rural community, a new industry moving to the community that will provide employment and capital for local people is a relatively permanent shift in the community environment.

In planning for educational change, the local school system must be sensitive to whether the shift in community environment is of a permanent or transitory, cyclical nature. If the shift is of a permanent nature and is significantly related to mathematics, it is important that related curricular change involve the entire staff that is teaching mathematics and that all levels of the curriculum be assessed in terms of the impact of the environmental shift. For the value or attitudinal shift that will wane, a school system can elect to do nothing until the community's pressure for change dissipates. Stonewalling until the problem goes away is not a viable option for a permanent shift in the community environment. The issues and problems become more extreme as time passes, and the need and force for change become more extreme and harder to react to in a sensible, calculated fashion. Thus, a major task of local school leaders is to judge whether a shift in the community environment is cyclical or permanent. A teacher cannot return to precomputer days; children need to acquire computer literacy in order to live full, productive lives. Therefore, a permanent shift in the state of the community environment defines an imperative for change.

The judgment of whether a change of state is of a cyclical or permanent nature is not always easy. Currently an issue of prime significance is basic skills and the associated issue of competency testing. It is easy to make a case that this issue represents a recurring attitude and domain of interest for concerned people in school communities. During the early part of this century, this was a major concern evident in the popular press. Induction tests for the draft in the early part of World War II reestablished this issue, and Bestor (1953) again focused attention on this arena of concern for basic skills. It seems apparent that the domain at issue is recurring and cyclical.

However, two new features have developed within this domain of concern. First, legislatures and school boards in previous periods of peak interest for basic skills have made decisions concerning opportunities to learn in the school program. For example, a legislative action in many states a number of years ago was to require that all high school graduates must have a Carnegie unit of mathematics but stipulated no quality or extent of learning. Legislation for minimal-competency testing goes beyond establishing the opportunity to learn to include the criteria of quality and extent of learning. It can be argued that this change, which bears the force of law, will reduce the cyclical nature of this issue.

The second new feature of this domain of concern is the change in the value associated with having a high school diploma. Green (1980) makes a compelling argument that attaining a high school diploma was viewed quite positively during the period when many individuals were not completing high school. However, since the mid-1960s when 75 percent of the seventeen-year-olds were graduating from high school, the values have shifted so that not graduating carries a stigma. In other words, whereas the graduate was formerly viewed positively, now the nongraduate is viewed negatively.

A corollary to Green's argument is that this necessarily shifts the focus of concern from the symbol of attainment of graduation, the diploma, to achievement. This analytic argument suggests that continuing concern for achievement and what it means not to have a diploma will be addressed through continued attention to minimal-competency testing and the achievement of basic skills. If Green's argument is accepted, it would be reasonable to expect that the cyclical pattern of attention to the issue of basic skills would be further reduced and that the planning for change by school leaders should expect a continuation of interest in basic skills. The inescapable conclusion is that if schools are concerned about the community's support, then they must demonstrate that basic skills are being dealt with effectively or that the school system is making changes to give more attention to basic skills. The interests and concerns about basic skills present in the community environment will not go away or diminish in significance.

The implications for the process of change and for the responsibilities of school personnel that derive from the shifts in the community environment are primarily in terms of identifying needed areas of change that will capitalize on the potential for the community's support for that change. If change is planned in terms of local support, then the likelihood is increased for administrative and school-board decisions in the interests of mathematics. A related and very significant outcome of tuning programs for change to shifts in the community environment is that the perception of the school by the community is enhanced, a payoff that should not be ignored in an era when many bond issues and school levies are rejected by voters.

Finally, being familiar with some history of mathematics education and education in general is helpful in determining if issues are of a cyclical or permanent nature. The advantage in decision making comes from knowing whether an issue or problem will be of continuing concern; if so, planning, implementation, and evaluation can be focused on long-term, continuing action and decision making instead of on the transitory problems associated with cyclical changes of state in the community environment.

Cumulative Inattention

A problem area may exist for years without being recognized; if it is recognized, effective action for change may not be taken. The actions needed to treat the problem can be particularly frustrating to those who recognize the problem because results are so hard to attain. Four scenarios are given below to indicate the types of frustrations encountered.

Scenario I. When a problem is recognized, expedient, stop-gap measures are taken because of the extreme need that has developed. The current shortage of qualified secondary school mathematics teachers in many areas of the United States is one example. Although evidence was available in the early 1970s that this problem was developing, it was not widely communicated; furthermore, the complex interaction of related economic and attitudinal factors was too vaguely understood to yield a comfortable solution. A stop-gap measure that has been taken by some school systems has been to shift the school organization from the 6–3–3 plan for elementary, junior high, and senior high school to a middle school arrangement that is organized x–x–4. Moving the ninth-grade students to the high school allows the middle school to be considered an elementary school for the purpose of assigning certified teachers to the school in many states. This solution also treats another problem that has developed with decreasing enrollments in the elementary grades, namely, the overstaffing of tenured elementary teachers. Although the middle school concept can be a powerful and effective organizational philosophy, moving to that structure does not treat the basic problem of a lack of qualified teachers for pupils of that age in mathematics (as well as some other fields). The change is expedient but is quite difficult to undo once made. Frequently a long-standing problem that has become so extreme that something must be done yields a change that does not treat the actual problem but only the superficial symptoms of the problem.

Scenario II. When a problem is recognized, individuals continue to talk of the problem but make no attempt to change what is done or how people think or act about the problem. For example, estimation has been recognized in many sets of curricular recommendations through the years as an important basic skill. The evidence indicates that few teachers devote much attention either to estimation in the sense of approximate computation or to estimation in measurement. Thus, students do not acquire a significant degree of proficiency in estimation. Interestingly, although the problem has been recognized for decades, the behavior of professionals has primarily been concerned with the reidentification of the problem with little investment in identifying strategies for changing how estimation is taught.

Scenario III. When a problem is recognized, nothing is done that proves effective. Unlike the previous situation, many different strategies for change are attempted, but each fails to have lasting, continuing impact. The curricular topics of probability and statistics and the efforts toward increasing the emphasis on these topics are examples. The need for increased attention to these topics has long been recognized by thoughtful mathematics educators. Significant attempts at communicating the need have been attempted and curricular materials have been prepared and tested, but little significant change is evident in the amount of curricular attention given to the topics.

This is perhaps the most frustrating of the scenarios. The problem has been recognized and communicated. Strategies for change have been attempted, but little evidence of change exists. Perhaps the traditions in the curriculum are too strong to overcome. Perhaps the principal actors in the implementation of this curricular change, the teachers, are not convinced of its significance or, alternatively, do not feel comfortable with their knowledge and intuitions about the subject matter. Perhaps among the problems and areas of need that require change, probability and statistics are simply of low priority for most people. New attempts at change will prove unsuccessful unless the previously used strategies for change are extended or replaced by strategies designed to deal with the fundamental characteristics of the resistance to the change.

Scenario IV. The problem has only recently been discovered, and there is insufficient experience or evidence from research to guide the development of a strategy for change. For example, consider the problem of basic skills and the problems students have in remembering. Most school mathematics programs teach the basic skills, and a large portion of the students can exhibit those skills at some point in the instructional sequence. Unfortunately, many of the students who can exhibit the skills immediately after instruction forget them by the time they must use them some weeks later either in application or in a testing situation. Interestingly, in a period of profound concern about basic skills, little attention is being given by researchers in mathematics education to how memory works or to what instructional characteristics facilitate memory. Duren (1980) conducted a comprehensive survey of literature concerned with memory and mathematics learning and found very few recent studies that address the problems associated with understanding or facilitating the processes of acquiring skill in remembering mathematics. In short, although memory is a fundamental cognitive process, the profession has ignored one of the most obvious possible means of improving children's performance in testing on basic skills in a period of profound concern for improving that performance. We do not know enough to design a rational strategy for change in this domain.

Each of these scenarios describes a situation in which the effects of inattention to a problem compromise making changes that would solve the problem or decrease a need. In each of the scenarios, one has the feeling that external wisdom is needed to help solve the problem. For most of the scenarios, if the problem or need were adequately communicated to outsiders, the outsiders could bring support and pressure to bear on the problem to contribute to its solution.

Private Institutions, Agencies, and Businesses

Businesses and groups that operate outside the control of schools or the government agencies that are responsible for education can make decisions that affect change in the mathematics curriculum in fundamental ways. For example, several commercial businesses whose revenues are derived primarily from education can make decisions affecting the selection of content for mathematics and the way that content is taught. Companies and nonprofit organizations that produce tests and texts are cases in point. A change of the selection of content for college-entrance testing can have a major impact on curricular decision making. The Thirty-second Yearbook of the National Council of Teachers of Mathematics (1970) documents the role of college-entrance testing in establishing the concept of function and graphs as an important portion of the secondary school curriculum. Prior to establishing an emphasis on this content in testing, few textbooks contained much material on these topics. After items on these topics began appearing in testing, the emphasis in texts dramatically increased.

Most firms that produce tests behave quite responsibly concerning building new content on important tests. They are careful to include mathematicians and mathematics educators in their planning for the future and to communicate changes in emphasis to the mathematics education community. No change, such as allowing students to use calculators while taking tests, is made without careful consultation with leaders in the field. Indeed, the resultant conservatism stemming from the awareness of the extreme role that they can play as determiners of curriculum is a major barrier for change in curricular emphasis in school mathematics. The firms that produce standardized tests must assure the users of the results of testing a stability of the characteristics of a test from one year to the next in order to yield useful data for interpretation.

Publishers of textbooks are caught in a similar role in determining curricula. Given the major markets, adoptions by states and large cities, and the extensive conditions concerning content that must be covered, special features that must be included in the text and supplementary materials, and the traditions that must be respected if a text is to qualify for adoption, the authors and editors have little choice to be innovative. Data on past sales indicate that the publishing house that brings out an extreme or highly innovative book takes a profound risk in the marketplace.

For publishers of texts and tests, the conditions for innovation and change are determined by the marketplace. It is easy to assign blame for the rigidity—the lack of change—to the publishers' apparent failure to support innovation and change. Perhaps the users and consumers of tests and teaching materials need to relax their conditions to make risk taking by the publishers more attractive. Such a relaxation is highly unlikely because in the strong traditions of local control, many different groups and individuals are responsible for the decisions.

Concluding Discussion

The effects of the external environment on changes at the local school level are significant and imposing. On the one hand, the external environment can force a change or provide an opportunity for a change. On the other hand, it can be a profound conservative force that is a barrier to educational change. One can analyze the process of change at the local level and in terms of how the local school community operates with some feeling that the system is rational. Indeed, the system appears sufficiently rational for it to be controlled to produce desired changes if the system is careful, thoughtful, and systematic in collecting and using information and in planning. In short, mathematics educators can feel optimistic about being agents with some control over the process of change.

The temptation in examining the external environment and its effect on the processes of local change is to accept the creed of the alcoholic: "Give me the courage to change the things I can change, the serenity to accept the things I cannot change, and the wisdom to know the difference." Although the comfort of this attitude is apparent, mere acceptance is not sufficient. The professional in mathematics education must be more aggressive in assessing when to take advantage of the external environment to produce desired changes. Few situations in which educational change is forced by the external environment can be dealt with by simple, serene acceptance. Typically, a change requires new knowledge and understandings, and if the external environment offers a situation conducive to change, then the professional at the local school level must be ready to use the situation to the advantage of learners and teachers of mathematics. The professional must be informed and ready to act in a knowledgeable and responsible fashion.

The most frustrating and puzzling aspect of the role of the external environment in the process of change is what can be done to influence it. In most facets of the external environment, the power and responsibility for decision making is dissipated and fragmented. If responsible individuals can be identified, they are often beyond other than indirect influence and pressure. Frequently, the change that must be made in the external environment boils down to convincing people to change their attitudes. However, the mechanisms for communicating information and convincing people of a need are not always readily accessible or identifiable. Local school personnel should be constantly on the lookout for the mechanisms and processes that will allow them to communicate their problems and needs to the external environment. They must realize that the local system cannot operate effectively in vacuo; they must look beyond the local school community in terms of building support for change.

The importance of the external environment as an element in the process of change will not decrease in the future. For most local school systems, it impels change through taking action that requires change, through providing resources for change that would be otherwise unavailable, and by providing the attitudinal and environmental shifts that indicate a need for change. Serenity is not enough.

REFERENCES

Bestor, Arthur E. *Educational Wastelands: The Retreat from Learning in Our Public Schools.* Urbana, Ill.: University of Illinois Press, 1953.

Conference Board of the Mathematical Sciences, National Advisory Committee on Mathematical Education. *Overview and Analysis of School Mathematics, Grades K–12*. Reston, Va.: National Council of Teachers of Mathematics, 1975.

Duren, Phillip Edward. "Teaching Reconstruction Memory Strategies to Seventh Grade Students in a Problem Solving Setting." Doctoral dissertation, Ohio State University, 1980.

Green, Thomas F. *Predicting the Behavior of the Educational System*. Syracuse, N.Y.: Syracuse University Press, 1980.

McLaughlin, Milbrey Wallin. *Evaluation and Reform: The Elementary and Secondary Education Act of 1965, Title I*. Cambridge, Mass.: Ballinger, 1975.

_____. "Implementation as Mutual Adaptations: Change in Classroom Organization." *Teachers College Record* 77 (February 1976): 339–51.

National Council of Teachers of Mathematics. *A History of Mathematics Education in the United States and Canada*. Thirty-second Yearbook of the NCTM, edited by Phillip S. Jones. Washington, D.C.: The Council, 1970.

Wilkes, Richard, Dorothy Strong, and Dwight Coblentz. "Guidelines for Designing the In-Service Program." In *An In-Service Handbook for Mathematics Education*, edited by Alan Osborne. Reston, Va.: National Council of Teachers of Mathematics, 1977.

12

The Courage to Change: Influencing the Decision Makers

JACK PRICE

IF THE NCTM's recommendations for mathematics education in the 1980s are to escape the fate of previous reform movements, mathematics educators must enlist the assistance of many groups of decision makers and opinion molders. As twenty-five years of stalled reform have shown, mathematics educators alone can do little to effect changes in curriculum, methods, or public acceptance. Decision makers and opinion molders, both inside and outside the profession, must be influenced to support planned change. These groups will then supply the impetus to make the change possible and effective.

Those groups outside the profession include the media, federal and state legislatures, textbook publishers, and test publishers. Inside the profession, those who must be encouraged to join NCTM in seeking reform include teacher-training institutions and other colleges and universities, state departments of education, district administration and local boards, and colleagues.

The purpose of this chapter is to define specific positive programs for enlisting the aid of these groups as each district and school attempts to implement NCTM's recommendations.

Decision Makers outside the Profession

The Media

The media can be a major ally of curriculum reform if they are carefully nurtured. In most smaller communities the daily or weekly newspaper exerts a force far beyond its apparent power. One should never underestimate this power. By the way in which ideas are presented, public opinion can be molded for or against a movement. In smaller cities, school principals and mathematics education leaders should be on a first-name basis with the editor of the local paper. In larger cities, they must cultivate the reporter who is assigned to the education beat and keep that person informed.

It is extremely important that representatives of the newspapers, in particular, be in on the ground floor of any plan to improve mathematics curriculum or instruction. Although it is probably not necessary to include the press in the planning sessions, they should be brought in as soon as plans are completed and ready to implement. Articles may not appear immediately and may not even be desired at this point, but through this contact the reporter receives valuable background information that can be used later to prepare an article or series of articles.

Radio and television stations may be less interested and less valuable as a resource at the beginning than the newspaper. However, as the plan goes into full swing or as evaluation results become available, these media may well be recruited to tell the story.

Many of the NCTM recommendations lend themselves to the media approach. The press can be enlisted in promoting such moves as making problem solving the focus of school mathematics, increasing the use of calculators and computers, increasing mathematics offerings, determining alternative methods for evaluating programs and performance, and increasing public support for education in general and mathematics education in particular.

For example, a reporter could be invited to a classroom demonstration of a good lesson in developing one or more problem-solving skills. With an understanding of the process, the reporter can better promote the local movement toward a focus on problem solving. Personal experience with calculators and computers is also important. If these instruments are available in the classroom, their power as teaching tools should be demonstrated to the press.

There is nothing more helpful than positive press. A positive article involving firsthand knowledge will do much to diminish some of the cries of outrage that will appear in the "Letters to the Editor" column in the aftermath of an article on declining test scores. An informed media campaign long before the test scores come out should help the public understand standardized testing and realize that not all children can be above the median and that some will even be below the mean. A series of training sessions for the board of education will help and will be doubly effective if the media also pick up on the sessions.

The press and the public must also come to realize that tests are not the only methods for evaluating programs. For example, how does the absentee rate in mathematics compare with that of the total school? How about discipline referrals? How successful is the job-placement program? How interested are parents? The psychological evaluation may be at times as valuable or more valuable than the cognitive examination of the program.

Finally, accurate information, carefully written and attractively presented either in print or on the airwaves, will do much to increase public support of education. To achieve good press from the local media, it is necessary to keep in touch with them and keep them informed. To paraphrase columnist Ann Landers, good public relations is like taking out the garbage—if you don't do it regularly, you have no right to complain when something begins to smell.

Federal and State Legislatures

If the media affect us somewhat indirectly, federal and state legislatures cover us like a blanket. Perhaps no other groups have had such a profound effect on education over the past decade. Legislators have created far more programs than any group of educational researchers and have spent more special-interest money on education than in any other endeavor. Further, legislators have probably had the most to do with making the NCTM recommendations necessary. To name but a few of their generally simplistic solutions to complex problems, they have locked in computational skills at the expense of problem solving, made a morass of the process of granting credentials, and reduced budgets to signal the end of supervisors and in-service education.

If educators are not pleased with the result, we have only ourselves to blame. In general, legislators respond to the loudest voices—those who command votes or campaign funds. Parents and other groups have co-opted the lobbying process. Education has rarely been able to get its act together enough to speak as one voice; administrators say one thing, teachers say another, boards say a third. When all can speak together, effective lobbying and beneficial legislation can result.

It is within the capability of the total mathematics education community to convince legislators that a wider range of mathematics programs should be available, that basic skills include more than just computation, that central office staff are important to curricular development and instructional improvement, and that funds for in-service education are needed and can be well spent. If it is possible, how is it to be done?

Personal contact is as important here as with the media. Legislators often have recess. Invite them for a school visit, particularly to classrooms, during one of these relatively uneventful periods. Good demonstrations can give legislators necessary background, but they cannot be expected to carry on from there alone as a reporter would. Schedule some time for face-to-face discussion on the NCTM recommendations, stressing those with which the district or school are most concerned. If possible, have a definite legislative proposal ready and make certain that broad-based support can be shown for the concept if not for the actual wording.

One of the most significant state-funded in-service programs ever attempted came about as a result of legislative interest by one legislator, the late California state senator George Miller. For four years, more than $0.5 million a year was spent in mathematics in-service training. Literally thousands of teachers and students benefited from this one well-researched program. Similar programs were subsequently established in the state of Washington.

It is not enough to plant ideas. The ground must be carefully tended. This means that the legislator's support staff must be contacted and kept as well informed as the lawmaker. It also means that as the bill moves through committees and the floor, support must be given and work must be done behind the scenes with the other legislators. And all through this, the support of teachers, administrators, board members, and the state department of education must be evident.

Legislators like to believe the bills they sponsor are popular. They count letters and telephone calls, and they categorize the support. Build a coalition across many different positions and ranks and make this known time and again. It works for welfare workers, truck drivers, and realtors, among many others. If we truly believe in the cause of mathematics education, we can do no less.

Textbook Publishers

Of all the outside interests, textbook publishers are probably the most willing to listen. If we don't buy, they are out of business. If there is any lasting effect from the curriculum reform efforts of the past twenty-five years, it is in commercially published texts. They little resemble the texts of 1960, and changes in content and method no longer require fifteen to twenty years to find their way into print.

Although we may not realize it, much of what we say to a publisher's representative about text material is transmitted to the home office. Books are written and revised based on the comments of users or potential users. Market research is a way of life for publishers. We can take advantage of that situation to influence needed change.

Publishers can provide assistance with the current recommendations by incorporating problem-solving activities into their materials and by making certain that more than drill and practice on fundamentals appears in their tests. Publishers should be encouraged to include material and activities on computers and calculators and to provide a wider range of choices for all students.

Local and state mathematics councils can play an important part in influencing publishers by including them in conferences and seminars not just as exhibitors but as participants. The leadership of the councils should be prepared to have round-table discussions with representatives from various publishers outlining the recommendations as well as the specific needs of the students of their areas.

Finally, textbook authors are our colleagues. They must be made aware of the necessity for the materials they write to reflect the recommendations of the NCTM. Although publishers and editors keep tabs on what sells and often hesitate to make major changes before a trend is well established, authors can influence their thinking through author-editor conferences and subtle revisions of existing materials.

Textbook publishers profited greatly from the reform movement of the 1960s and are amenable to change. New methods in printing and composing allow them to speed up the technical process, and well-informed authors can speed the writing process. If publishers know the needs, they can produce far more rapidly than in the past. We need to keep them informed through their sales representatives, their editors, and their authors. Changes can be made and made quickly, but publishers have to know what we want changed.

Test Publishers

A respected mathematics educator has commented on test publishers:

> Given the amorphous character of the populations that use and construct the tests, I am not sure that it is possible to have a strong effect on the marketplace of testing. . . . It seems to be bound by traditions and to be very resistive of influence.

The major commercial test publishers do appear resistant to influence, and the Nader-like approaches and "truth in testing" laws have done little to move them. Yet they could be of enormous assistance, particularly with the recommendations involving problem solving and the use of other than standardized tests in the evaluation of programs.

Teachers tend to teach to tests, and they should if those tests cover their objectives. What better way to be certain that problem solving is taught than to test for it?

Certainly there is much more opportunity for successful influence at the state or federal level when public funds are involved. For example, the National Assessment of Educational Progress has been careful to include mathematics educators as it develops its assessment programs. Since NCTM committee members have access to data before their release to the public, they can influence the manner in which the data are presented, making certain the data are accurately reported.

When states develop their own testing program, mathematics educators can have—and have had—significant influence. In California, for example, SCIMA, a test developed through state funds, was based on the state mathematics framework and encompassed four levels of cognitive development. It was truly an assessment tool, since various scales were randomly assigned to students in the state. It was not an individual assessment instrument. The test was developed by a group of mathematics educators who covered the spectrum from college professors to classroom teachers. It served as the forerunner of the present California Assessment Program.

Other states, notably Michigan, Oregon, and Wisconsin, have had similar experiences in influencing test construction. They have shown that it is possible to develop tests that meet the objectives of the program, that treat various levels in the taxonomy of cognition, and that are both professionally and statistically honest. The national commercial test publishers could use the same set of criteria if they wished.

It is difficult not to dismiss as fruitless any effort to influence these publishers. But mathematics educators must make an effort to exert an influence. We must continue to invite publishers of tests to conferences and offer them opportunities for face-to-face discussion. If change is to be effected, it will have to occur at the national level.

Decision Makers within the Profession

Colleges and Universities

There is a general hypothesis that the amount of change a teacher will willingly undergo is inversely proportional to the teacher's grade level. The higher the grade, the less likely the teacher is to change. Because colleges and universities prepare our teachers, conduct most of the research, and establish in-service training programs, a special effort must be made to make them a counterexample for this hypothesis. Mathematics education cannot afford to have school districts moving one way while colleges move another or for one or the other to refuse to move. Once again, communication is important. College and university personnel, if they know about the trends, can modify their programs to ensure that subject matter and methods change to meet new standards. By sending out teachers who are better prepared, they raise esteem for the profession and, as a consequence, gain more public support for educators in general.

School district personnel should create opportunities for college professors to deal directly with teachers in the teachers' own setting. A frank discussion of problems and a careful delineation of teachers' needs can result in upgraded preservice and in-service programs. Programs are easy to develop alone, but unless teachers or students have a stake in them, these programs are doomed to wither and die.

In exchange for information and access, college personnel should give districts a piece of the action. Often teachers are well aware of their weaknesses. They simply need someone to help them find their strengths. This can be done through classroom visits, conferences, and just listening.

This is not to imply that teachers and supervisors have all the answers. Local college personnel may be light years ahead. But the worst way to involve school district personnel is to drag them kicking and screaming into programs that have been "designed to help them" without allowing them to participate at the planning stages. Even the most ardent defender of the reform movement of the 1960s will agree that teachers were not brought along. Materials were developed for them, and they were told how to teach them. This lack of involvement at the developmental stages probably contributed more than any other single factor to the lack of acceptance and the lack of implementation of the reforms in the classroom.

We cannot afford another abortive attempt to change. Regardless of who the leaders are in any area, they must make certain that all levels of mathematics educators are involved. *Who* takes the lead does not matter as much as the fact that someone does and that the local school is the focus.

State Departments of Education

When a state has a well-functioning department of education, its staff should be enlisted in the battle for improving mathematics education. Because of their proximity to the seat of power in the capitol, state departments can influence legislators as well as local districts. They are in an ideal position to move recommendations for reform on a state level. The Miller Mathematics Improvement Programs, alluded to earlier, were a direct result of the intervention of a state consultant in mathematics. The same can hold in any push to implement the agenda for the 1980s.

State department officials can exert pressure in three major areas—curricular reform, credentials reform, and public support. One state department consultant has already submitted to a state board of education a revised state framework in mathematics that moves problem solving from only a curricular strand to the focus for all mathematics instruction. Because the state criteria for text selection are based on the state framework, the pressure will be on textbook publishers to come through.

State consultants cannot help, however, if they are not informed. Although most of them are active at the national level, they have lost some touch with the daily classroom. It is the particular responsibility of local leadership to narrow this gap. Certainly the state consultant should take the first step, but it is often difficult for a staff person to push his or her way into a local school situation. Thus, it is necessary for local leadership to schedule periodic meetings for the consultant to talk with teachers, principals, central office staff, and education professors. These meetings can explore the recommendations in depth, assess the present status, and design plans to move from what *is* to what should be. It is in meetings like these that a major legislative proposal or significant reform effort can have its genesis.

There is no substitute for information. Therefore, local district staff or leaders of local mathematics councils should take the initiative if a void exists at the state level. Whatever happens must happen as a result of cooperative effort among teachers, supervisors, college and university personnel, and the state department of education. As most management textbooks will say, a great deal can be accomplished if there is no concern about who gets the credit (generally everyone will).

District Administration and Local Boards

For change to be made at the local level, the support and encouragement of the district administration and the board of education must be obtained. Often boards and superintendents are overwhelmed with budgets, personnel problems, transportation, extracurricular activities, and similar considerations. Seldom, if ever, do they have an opportunity to concern themselves with curriculum and instruction. The emphasis should be turned around.

One method to provide information is a report to the board on the mathematics program and the plans for improvement. When accompanied by a classroom visit, this can be a powerful means of driving home a message about the mathematics program. Involving board members in mathematics activities is another, less obvious means of eliciting support. Their memory of attendance at a mathematics field day, a curriculum conference, or an awards ceremony for young people of the district can persuade them to support requests made later. Board members and superintendents asked to preside at carefully selected section meetings of a mathematics conference can be relied on later to support requests for program change.

In-service training is undoubtedly a major component of any movement to improve mathematics education. Thus, staff development should be recognized somewhere in the teachers' collective bargaining contract. Funding and released time are areas in which both district and teachers may find ready agreement. Certainly the fact that both teachers and central staff find it valuable will contribute greatly to the awareness level of both superintendent and board.

It is most important to increase their awareness, through conferences, visits, negotiations, and even subliminal suggestions. New ideas should be repeated often prior to specific requests. Overt attempts to have both board and superintendent informed before they are called on to make a decision will be time well spent, and chances for success will increase in proportion to their awareness.

Colleagues

Perhaps the most important group of people to influence with respect to the NCTM recommendations for the 1980s are our colleagues. In the final analysis, they were responsible for not implementing the last wave of reform. Eighty percent of them do not belong to the NCTM and have little access to its recommendations or to the means by which they can be implemented. Yet, if success is to come, it must be through efforts at the grass roots.

How can we convince our colleagues to join the battle? The greatest encouragement is by example. Ideas for emphasizing problem solving or using calculators or improving efficiency in teaching can be imparted subtly and unobtrusively through sharing materials, invitations to visit the classroom, and informal give-and-take sessions in the office. Teachers who attend conferences and other in-service activities can influence other teachers to attend also. "Take a colleague" could well be the rallying cry.

At the lowest level, we could make certain that the materials selected reflect the program we wish to have. Often with some of our colleagues the textbook is the course of study. Careful selection of the right materials and skillful assistance in their effective use can move the program forward, perhaps more slowly than we would wish, but at least move it.

It seems obvious that at the high school level those on the periphery should also be included in any concerted effort to move the mathematics curriculum forward. The mathematics department must shed its purist and isolationist attitude and draw the circle larger. Teachers of business and shop or vocational mathematics should surely be encouraged to adopt new methods and materials that are more closely aligned with the NCTM recommendations. Further, the business department and the science department should become full partners in the drive for computer literacy. Problem solving and applications require career considerations, and mathematics must be seen as a tool in achieving the objective of individual economic and social self-sufficiency.

In the long run, our colleagues may present us with our most difficult selling job. But if we can convince them that we can truly make a change for the better, then we will have accomplished the last two recommendations. The public will see that we treat mathematics instruction with the professionalism it deserves. In return, the community will be encouraged to make the rewards for teaching commensurate with the responsibility. It is a circle: professionalism breeds respect; respect brings rewards; rewards improve professionalism.

The circle must begin somewhere. We, and our colleagues, can take the first step. No one—parent, teacher, administrator, student—should be satisfied with less than the best effort. The agenda for action for the 1980s is not complex. It takes only a little will, a little effort, a little more caring from each of us. If we can put forth the effort, if we can enlist the aid of those interests both outside and inside the mathematics profession, then we shall truly provide the best mathematics education we can give.

13

Implementing the "Agenda for Action"

MAX A. SOBEL

I MAGINE driving along a highway and spotting this attention-getter on a billboard:

Due to a lack of interest
Tomorrow has been canceled.

No amount of discussion about the nature of the process of change will have any impact if there is not genuine interest and support from those who are most influential in any curriculum change—the classroom teachers. Our experience with the "new math" taught us that curricular innovation can be advocated by leading educators, that evidence of change can be found in textbooks, that the need for change can be presented at professional conferences and yet that change will *not* take place if the classroom teacher is neither convinced about the need for change nor ready for new developments.

In the final two decades of this twentieth century, the mathematics curriculum presents a challenge that we simply cannot afford to let pass. We have stumbled through recommended programs of mathematics in the 1960s that often proved disastrous in their implementation. The back-to-basics movement of the 1970s brought the mathematics teaching profession into disrepute as our students gave evidence of their lack of ability to apply fundamental skills to the solution of real-life problems. In each case, the media was quick to react in a negative manner. Headlines such as the following were common:

"Computing the Minuses of Math Education"
"Failing Report Card Delivered on School Mathematics by Teachers"
"Back to Basics a Failure"

It was in this climate that the National Council of Teachers of Mathematics published its recommendations for school mathematics of the 1980s in *An Agenda for Action,* a document based on exhaustive studies. The professional reference book you are now reading has presented detailed commentaries on the nature of the change process. Now we are ready to proceed with the curricular changes that should enable the next generation of youth to meet the technological challenges of the twenty-first century.

By now it should be apparent that change is a complex process that comes about in slow stages. Indeed, if one looks at the changes that have taken place in the past hundred years in mathematics education and projects these forward by extrapolation for the coming hundred years, it seems safe to predict that progress will come more at the pace of a tortoise than a hare when we consider where we stand in the whole picture of the creation of our world.

Carl Sagan in *The Dragons of Eden* paints a vivid picture that is quite humbling in its presentation. He suggests that we imagine the 15-billion-year lifetime of the universe since the "big bang" compressed into a single year. Thus every billion years would correspond to about 24 days of this one year, and about 475 days would correspond to a single second. Under such a scheme, we would imagine 1 January of this year as the date of the creation of the universe and would then have to wait until about 25 September for the origin of life on earth. It is not until 21 December that the first insects appear on earth and Christmas Eve before the first dinosaurs appear. Continuing on, it is not until 10:30 P.M. of New Year's Eve that the first human beings appear on earth. At 11:00 P.M. we find the first use of stone tools, and the alphabet is invented at fifty-one seconds into the last minute of the year. Zero and decimals are created at 11:59:57 P.M. of this last day! Indeed, on Sagan's scale, *all of recorded history occupies the last ten seconds of this final day of the year.*

Where are we now? This decade of increased attention to science and technology, the start of our exploration of outer space, and NCTM's *Agenda for Action* all occupy the very first second of New Year's Day of the second year of the existence of the universe. Thus it is obvious that we occupy but a pause in this cosmic calendar, and therefore it is imperative that we make the most of it!

If we are to be successful at hastening progress and implementing the desired changes in this decade, then we must begin by first accentuating the positive aspects of our present programs. We must inform the general public and the media of the many good things that are happening daily in our classrooms. We need to encourage public support by involving the public in the entire educational process. It is not enough to contact parents when their child misbehaves. We need to extend ourselves and call home when something special happens in class—when a student makes an exciting discovery, comes up with an original proof, gets an A on a test, or even passes a test after repeated failures!

A significant part of the change process lies in educating the public with respect to the important role that the classroom teacher of mathematics plays in the education of their children. This position is emphasized as one of the basic recommendations of the *Agenda:*

> Public support for mathematics instruction [must] be raised to a level commensurate with the importance of mathematical understanding to individuals and society.

The time has come to force the public to realize that we are dealing with the most precious commodity in our society—their children. We must demand that public support be raised to a level commensurate with the importance of their children! Until we can achieve this goal, there can be little expectation of success for any new programs in our schools.

There are a number of challenges that must be faced if the recommendations in the *Agenda* are to be fulfilled and if change is to come about in this decade. Let us explore several of these.

Teacher Preparation

The preparation of elementary school teachers of mathematics remains inadequate for contemporary programs. A survey conducted by NCTM revealed that 25 percent of all prospective elementary school teachers still take three credit hours or less of mathematics during their undergraduate programs. Especially disturbing is the fact that about 8 percent take *no* mathematics at all, and yet they are certified to teach and do so! There is a desperate need to improve certification requirements to demand more work in undergraduate mathematics and in the teaching of elementary mathematics. Little change can be expected if teachers are not adequately trained.

Equally disturbing is the content of education programs for preservice elementary school teachers. Considering the NCTM recommendations for the curriculum of the 1980s, it is obvious that changes must be made in such courses when surveys indicate that only 11 percent of the programs emphasize the role and use of calculators and only 5 percent list computer mathematics as a topic of emphasis. As one report by the NCTM Commission on the Education of Teachers of Mathematics (CETM) points out, although the content of such courses has changed with time, the question of their adequacy for the preparation of mathematics teachers for the future still looms large.

It seems clear, according to the report, that the content of preservice courses is not commensurate with the needs of the elementary classroom teacher of today, let alone in coming decades. Furthermore, the offerings in mathematics education for this group of teachers appear totally inadequate for meeting current needs. Teacher-training institutions must begin immediately to address themselves to this issue.

The preparation of prospective secondary school teachers of mathematics needs careful rethinking as well. It appears that the time has come to move into a five-year program that includes a full semester, if not a full year, of internship under careful and continued supervision. In a sense parents are entrusting the lives of their children to the classroom teacher, and we need careful training in mathematics, pedagogy, psychology, psychotherapy, and whatever else it may take to help to shape the lives of our future citizens.

In-Service Education

Past NCTM surveys conducted by the CETM clearly show that most teachers want in-service programs provided such programs are meaningful and designed to meet their expressed needs. At the top of the list of topics requested by teachers is that of techniques of motivation.

The recommendations set forth in the *Agenda for Action* will have scant hope of success—and the change process scant chance of being effectively initiated—without extensive work in such areas as techniques of motivation, applications of mathematics at all grade levels, effective use of calculators, the role of the computer in the mathematics classroom, and so forth. Such in-service training is desperately needed at all levels of instruction, and we must demand local funds to support such efforts if federal funding is not available. Curricular change without adequate teacher preparation for such change is doomed to failure from the outset.

The NCTM's publication *An In-Service Handbook for Mathematics Education* provides guidelines for designing and implementing in-service programs and warrants careful study by those in a position to initiate such programs.

Teacher groups should make a concerted effort to demand time and financial assistance from local school boards for in-service training on both mathematical and pedagogical topics of current concern. At the same time, teacher unions and associations, together with the NCTM, must exert pressure on our legislators to provide for in-service training to be sponsored by the National Science Foundation (NSF) and other federal agencies.

The NCTM provides leadership training by sponsoring conferences on in-service programs that can then be replicated in individual communities. For example, at a conference on the use of calculators in mathematics instruction, leaders were presented with programs and materials for use in conducting in-service institutes on this topic. State professional associations could do likewise for the leaders in their state who could then serve as mathematical field agents for the dissemination of new ideas and approaches.

Professionalism

One of the recommendations for school mathematics of the 1980s is that "mathematics teachers demand of themselves and their colleagues a high level of professionalism." We are incensed when we hear of lawyers or physicians who behave in an unprofessional way and yet receive the protection of their colleagues. How often do we do the same, if only by our silence? The *Agenda for Action* clearly states the professional position on this matter:

> Teachers as a profession should insist that all members maintain a consistently high standard of professional behavior. The profession is not obligated to protect those individuals who refuse to live up to reasonable professional standards.

The time has come for the teaching profession to take dramatic steps toward self-governance if we are to ensure the success of curricular change. A professional code of ethics should be prepared by and for the members of our Council that could serve as a model for the entire profession.

Membership in professional organizations is essential if the teaching profession is to be heard and if we are to reach those who will be actively involved in the change process. Thus it is especially disturbing to find that membership in the NCTM has shown a slow but steady decline throughout the latter half of the 1970s and into the 1980s as well. It is even more disturbing to note that the vast majority of teachers of mathematics are *not* members of this organization. The total number of teachers of mathematics in the United States and Canada is approximately 210 000 at the secondary level, 1.5 million at the elementary level, and 28 000 at the collegiate level. Considering that membership in NCTM, including institutional membership, is approximately 65 000, it is obvious that we are failing to reach a large number of those involved in the daily task of teaching mathematics to our children.

Every teacher of mathematics should feel a personal obligation to try to reach some of their unaligned colleagues and attempt to convince them of the benefits and necessity of working together in implementing desired change in our curriculum. Change will be difficult, if not impossible, unless we are able to work cooperatively with the vast majority of those involved in the daily task of teaching mathematics.

The benefits of NCTM membership should be emphasized to those who do not belong. Our professional journals and publications provide up-to-date information and ideas both in mathematics and pedagogy at all levels. Regional meetings are scheduled throughout the United States and Canada so that one is within a 500-mile radius of each member every two years. And our collective voice is heard as we send representatives to numerous other organizations and federal agencies in an effort to improve the status and support of mathematics education.

The Teacher Shortage

At this writing there exists a significant shortage of teachers of secondary mathematics in the United States. Although the situation may change, all evidence seems to indicate that this shortage will continue and worsen alarmingly throughout the 1980s. The problem is especially pronounced in large cities where we so often look for leadership in the change process. Some cities have reported less than one applicant for each ten teaching positions available in secondary mathematics!

It is axiomatic that change cannot take place without fully qualified teachers in the classroom. Yet the most popular strategy for dealing with a shortage of qualified applicants has been to assign teachers outside their major field of preparation to teach mathematics. Other reported strategies include increasing class size, using teachers with emergency certificates, lowering certification requirements, eliminating courses, and using substitutes.

The financial lure of positions in such fields as computer science has drained our supply of prospective teachers of mathematics. Schools of education report small numbers of students entering the field of secondary education, with mathematics suffering far more than other disciplines because of the availability of higher paying jobs in related areas. If this trend continues throughout this decade, we shall be faced with the impossible task of attempting to provide the appropriate training and education for life in a technological age with teachers who are not qualified to teach. In such a setting, teaching the standard program of mathematics is a monumental task; attempting to implement change in the curriculum becomes totally impossible.

Although it is incumbent on society to provide the appropriate incentives for attracting our best youth into the teaching profession, there are steps that the classroom teacher can and must take in the interim. As a first step, we must publicize the need for teachers of mathematics. Too many of our students fail to enter the teaching profession because they have heard that there is an overabundance of teachers. We must make it clear that this is not true in secondary mathematics.

Although NCTM can and will continue to dramatize the need for mathematics teachers, each of us should begin to take steps to encourage our *best* students to consider teaching as a career. There is no doubt that many of us have been influenced in the past by some teacher who helped steer our lives in a certain direction. What better evidence of our own love for teaching can we give than to sponsor an outstanding student and encourage that individual to become a teacher of mathematics!

Let us *show* our students what a wonderful life the teacher of mathematics enjoys. Let us be living examples of the joys of working with young people. Some of our students might consider joining the teaching profession if they were to see daily evidence of the genuine joys and satisfactions of teaching.

The Curriculum and the Teacher

The shortage of qualified teachers of mathematics comes at a time in our history when we can ill afford to decrease the quality or the quantity of mathematics offerings in our schools. We are faced with an unprecedented challenge from the Soviet Union that will require all our resources if we are to avoid becoming a second-rate nation with respect to technology during the remaining years of this century. Izaak Wirszup of the University of Chicago has made us aware of the dramatic expansion of the Soviet Union's educational program during the past decade. He notes that whereas fewer than 2 million students graduated from their secondary schools prior to 1960, this number jumped to over 5 million by 1980. This represents a success rate of approximately 98 percent, compared to 75 percent of all seventeen-year-olds graduating in the United States. Furthermore, adds Wirszup, all these graduates from Soviet secondary educational institutions have completed a two-year study of calculus!

In a report to the National Science Foundation, Wirszup wrote the following paragraph, which he underlined for emphasis:

> The Soviet Union's tremendous investment in human resources, unprecedented achievements in the education of the general population, and immense manpower pool in science and technology will have an immeasurable impact on that country's scientific, industrial and military strength. It is my considered opinion that the recent Soviet educational mobilization, although not as spectacular as the launching of the first Sputnik, poses a formidable challenge to the national security of the United States, one that is far more threatening than any in the past and one that will be much more difficult to meet.

A 1980 report on the status of science and engineering in the United States, prepared by the Department of Education and the National Science Foundation for the president, states that "the declining emphasis on science and mathematics is in marked contrast to [that of] other industrialized countries." The report continues:

> The current trend toward virtual scientific and technological illiteracy, unless reversed, means that important national decisions involving science and technology will be made increasingly on the basis of ignorance and misunderstanding.

Possibly the general public needs the shock of the Wirszup report to lend its full support to this recommendation from the *Agenda*:

> School districts should increase the amount of time students spend in the study of mathematics.

It appears obvious that we cannot meet the challenge presented by other industrialized nations without increasing and strengthening the mathematical offerings in our schools. And it is equally obvious that we are not able to proceed with such adjustments to our curriculum without an adequate supply of qualified teachers of mathematics. Probably one of the most disturbing causes of the teacher shortage cited by the NSF report is the "widespread indifference to the status of teachers." This public indifference *must* be changed if we are to meet the technological challenges of the remaining years of this decade. Each of us in the profession must accept the challenge of making the public aware of the dangers of indifference.

NCTM has been spreading the word about the curricular changes needed for the 1980s as well as the need for support of qualified teachers. Speakers have been sent to the annual meetings of such groups as the American Association of School Administrators, the National Association of Secondary School Principals, the National School Boards Association, and the Association for Supervision and Curriculum Development. Within each state and province of the United States and Canada we must encourage local mathematics associations to address these important issues with other groups to force realization that this is a societal issue for which all are responsible.

The Students

As we consider the mathematics program for the remainder of this century and as we consider the process of change, we must never lose sight of our students. Regardless of curricular changes, the basic characteristics and needs of our students will remain unchanged. These needs are the ones that young people (and indeed *all* individuals) have always had:

- The need for security
- The need for acceptance
- The need for affection
- The need for success
- The need to feel important

We must understand these basic needs as we provide special programs for all youth—the gifted and the slow, the underrepresented, the handicapped and the minority student, the college bound and the average. It is essential that we recognize and provide for them all.

This is our basic challenge: to meet the needs of the children we teach. Psychologist Haim Ginot addressed this point quite eloquently:

> I already know what a child needs. I know it by heart. They need to be accepted, respected, liked, and trusted; encouraged, supported, activated, and amused; able to explore, experiment, and achieve. Damn it! They need too much. All I lack is Solomon's wisdom, Freud's insight, Einstein's knowledge, and Florence Nightingale's dedication.

These are precisely the qualities we need for our difficult tasks—Solomon's wisdom, Freud's insight, Einstein's knowledge, and Nightingale's dedication. Nothing less will suffice! The effort to change and improve school mathematics must be an all-out commitment. When we consider the short second that we have in the cosmic universe we recognize that there really is not much time left for us to achieve the goals that we have set forth for ourselves.

Conclusion

Who is to carry out the changes? It is all too easy to leave the job to "them." *They* should provide a new curriculum, and *they* should implement change. This is easy to say and in keeping with the statistic quoted by one national leadership group that 50 percent of the members of any association are indifferent and may be considered "checkbook" members only. It is essential to realize that, as Pogo might say, *they* is *us*. No change is possible except by a joint effort of the entire educational community, with every teacher taking an active role in the process.

The spotlight that focused on the teacher of mathematics during the 1960s dimmed during the soul-searching years of the 1970s. It is time to spotlight the teacher of mathematics again—to show the general public how important mathematics teachers are to the future of their children and to the future of all society in this technological age.

The major ingredient in the process of change is the classroom teacher. Despite all other problems, the classroom teacher will spell the difference between success and failure in implementing our goals. Over sixty years ago J. W. Young (*The Teaching of Mathematics in the Elementary and Secondary School*) suggested that teachers ask themselves the following questions daily:

> Has each pupil profited by my presence in the class-room to-day? ... Is it possible for any pupil to say that he came to my class ready and willing to learn but that his teacher gave him no help? Could the class have obtained all that they got in the class hour to-day equally well from a lifeless book, from one another, or from private study?

Those who must answer no to the first or yes to either of the other questions are failing as teachers of mathematics. All of recorded history makes it quite clear that the most important feature in the learning process always has been and always will be the teacher. The classroom teacher is the key to learning, and no agenda can be effective unless our society is willing to ensure a supply of highly motivated, well-trained, competent teachers of mathematics at all grade levels.

The general public expects a lot from us! They want us to serve in many roles for their children: psychologist, psychotherapist, lawyer and judge, clergy, police, physician, nurse, and more. Furthermore, we are to teach their children ethics and morals from 8:30 to 3:30 so that they need not worry about such matters for the remaining seventeen hours of the day. And, oh yes, they expect us to teach their children mathematics as well! This is a tall order but one that good teachers have always accepted with pride and determination. The question that we need to raise now is what they, the public, will do for us. And then we must reemphasize the recommended action stated in the *Agenda:*

> *Society must provide the incentives that will attract and retain competent, fully prepared, qualified mathematics teachers.*

Change is inevitable, but unless the change process is managed with care and intelligence, we shall waste more years as we fumble for the appropriate course of action. Times are such that we do not have these years to waste. Each teacher of mathematics must become actively involved in a careful evaluation of existing programs and in the detailed implementation of new approaches. This will take a certain degree of serenity, a great deal of courage, and an almost infinite amount of wisdom.

Appendix

An Agenda for Action

Recommendations for School Mathematics of the 1980s

The National Council of Teachers of Mathematics recommends that—

1. problem solving be the focus of school mathematics in the 1980s;

2. basic skills in mathematics be defined to encompass more than computational facility;

3. mathematics programs take full advantage of the power of calculators and computers at all grade levels;

4. stringent standards of both effectiveness and efficiency be applied to the teaching of mathematics;

5. the success of mathematics programs and student learning be evaluated by a wider range of measures than conventional testing;

6. more mathematics study be required for all students and a flexible curriculum with a greater range of options be designed to accommodate the diverse needs of the student population;

7. mathematics teachers demand of themselves and their colleagues a high level of professionalism;

8. public support for mathematics instruction be raised to a level commensurate with the importance of mathematical understanding to individuals and society.

Recommendation 1
PROBLEM SOLVING MUST BE THE FOCUS
OF SCHOOL MATHEMATICS IN THE 1980s

The development of problem-solving ability should direct the efforts of mathematics educators through the next decade. Performance in problem solving will measure the effectiveness of our personal and national possession of mathematical competence.

Problem solving encompasses a multitude of routine and commonplace as well as nonroutine functions considered to be essential to the day-to-day living of every citizen. But it must also prepare individuals to deal with the special problems they will face in their individual careers.

Problem solving involves applying mathematics to the real world, serving the theory and practice of current and emerging sciences, and resolving issues that extend the frontiers of the mathematical sciences themselves.

This recommendation should not be interpreted to mean that the mathematics to be taught is solely a function of the particular mathematics needed at a given time to solve a given problem. Structural unity and the interrelationships of the whole should not be sacrificed.

True problem-solving power requires a wide repertoire of knowledge, not only of particular skills and concepts but also of the relationships among them and the fundamental principles that unify them. Each problem cannot be treated as an isolated example. This recommendation looks toward the need to solve problems in an uncertain future as well as here and now. As such, mathematics needs to be taught as mathematics, not as an adjunct to its fields of application. This demands a continuing attention to its internal cohesiveness and organizing principles as well as to its uses.

Recommended Actions

1.1 *The mathematics curriculum should be organized around problem solving.*

- The current organization of the curriculum emphasizes component computational skills apart from their application. These skills are necessary tools but should not determine the scope and sequence of the curriculum. The need of the student to deal with the personal, professional, and daily experiences of life requires a curriculum that emphasizes the selection and use of these skills in unexpected, unplanned settings.

- Mathematics programs of the 1980s must be designed to equip students with the mathematical methods that support the full range of problem solving, including—

    —the traditional concepts and techniques of computation and applications of mathematics to solve real-world problems, the rational and real number systems, the notion of function, the

202

use of mathematical symbolism to describe real-world relationships, the use of deductive and inductive reasoning to draw conclusions about such relationships, and the geometrical notions so useful in representing them;

—methods of gathering, organizing, and interpreting information, drawing and testing inferences from data, and communicating results;

—the use of the problem-solving capacities of computers to extend traditional problem-solving approaches and to implement new strategies of interaction and simulation;

—the use of imagery, visualization, and spatial concepts.

- Mathematics programs should give students experience in the application of mathematics, in selecting and matching strategies to the situation at hand. Students must learn to—

—formulate key questions;

—analyze and conceptualize problems;

—define the problem and the goal;

—discover patterns and similarities;

—seek out appropriate data;

—experiment;

—transfer skills and strategies to new situations;

—draw on background knowledge to apply mathematics.

- Fundamental to the development of problem-solving ability is an open mind, an attitude of curiosity and exploration, the willingness to probe, to try, to make intelligent guesses.

- The curriculum should maintain a balance between attention to the applications of mathematics and to fundamental concepts.

1.2 *The definition and language of problem solving in mathematics should be developed and expanded to include a broad range of strategies, processes, and modes of presentation that encompass the full potential of mathematical applications.*

- Computational activities in isolation from a context of application should not be labeled "problem solving."

- The definition of problem solving should not be limited to the conventional "word problem" mode.

- As new technology makes it possible, problems should be presented in more natural settings or in simulations of realistic conditions.

- Educators should give priority to the identification and analysis of specific problem-solving strategies.

- Educators should develop and disseminate examples of "good problems" and strategies and suggest the scope of problem-solving activities for each school level.

1.3 *Mathematics teachers should create classroom environments in which problem solving can flourish.*

- Students should be encouraged to question, experiment, estimate, explore, and suggest explanations. Problem solving, which is essentially a creative activity, cannot be built exclusively on routines, recipes, and formulas.

- The mathematics teacher should assist the student to read and understand problems presented in written form, to hear and understand problems presented orally, and to communicate about problems in a variety of modes and media.

- The mathematics curriculum should provide opportunities for the student to confront problem situations in a greater variety of forms than the traditional verbal formats alone; for example, presentation through activities, graphic models, observation of phenomena, schematic diagrams, simulation of realistic situations, and interaction with computer programs.

1.4 *Appropriate curricular materials to teach problem solving should be developed for all grade levels.*

- Most current materials strongly emphasize an algorithmic approach to the learning of mathematics, and as such they are inadequate to support or implement fully a problem-solving approach. Present textbook problems tend to be easily categorized and stylized and often bear little resemblance to highly diversified, real-life problems. They do not permit the full range of strategies and abilities actually demanded in realistic problem contexts.

- The potential of computing technology for increasing problem-solving ability should be explored and exploited by the development of creative and imaginative software.

1.5 *Mathematics programs of the 1980s should involve students in problem solving by presenting applications at all grade levels.*

- Applications should be presented that use the student's growing and changing repertoire of basic skills to solve a multitude of routine and commonplace problems essential to the day-to-day living of every citizen.

- Applications of mathematics to other disciplines such as the social sciences, business, engineering, and the natural sciences should be presented.

- The enormous versatility of mathematics should be illustrated by presenting as diversified a collection of applications as possible at the given grade level.

- At the college level, courses in mathematics and the mathematical

204

sciences should give prospective teachers experiences that develop their capacities in modeling and problem solving.

1.6 *Researchers and funding agencies should give priority in the 1980s to investigations into the nature of problem solving and to effective ways to develop problem solvers.*

- Support should be provided for—
  —the analysis of effective strategies;
  —the identification of effective techniques for teaching;
  —new programs aimed at preparing teachers for teaching problem-solving skills;
  —investigations of attitudes related to problem-solving skills;
  —the development of good prototype material for teaching the skills of problem solving, using all media.

Recommendation 2
THE CONCEPT OF BASIC SKILLS IN MATHEMATICS MUST ENCOMPASS MORE THAN COMPUTATIONAL FACILITY

There must be an acceptance of the full spectrum of basic skills and recognition that there is a wide variety of such skills beyond the mere computational if we are to design a basic skills component of the curriculum that enhances rather than undermines education.

We recognize as valid and genuine the concern expressed by many segments of society that basic skills be a part of the education of every child. However, the full scope of what is basic must include those things that are essential to meaningful and productive citizenship, both immediate and future.

The agreement among parents, educators, and mathematicians on the need for teaching basic skills with greater effectiveness unfortunately does not yet extend to a common understanding and acceptance of exactly what these basic skills should comprise.

Some groups narrowly limit them to routine computation at the expense of understanding, applications, and problem solving. This would leave little hope of developing the functionally *competent* student that all desire.

It must also be recognized that individual capacities, interests, and future directions might call for different emphases and different selections in matching basic skills to individual needs.

The time and energy that teachers and programs should be devoting to building beyond minimal foundations are sometimes skirted, being considered risky deviations from the minimal targets on which educators believe they will be judged. There is great pressure today to use all such time, energy, and re-

205

sources on overkill in the minimal target areas even though little added productivity may be achieved.

Rather than fostering a return to some acceptable common threshold of performance, the back-to-basics movement tends to place a low ceiling on mathematical competence—and this at the onset of an era in which daily life will be more deeply permeated by multiple and diverse uses of mathematics than ever before. Under these circumstances, even if improvement in rote computation takes place, a citizen who cannot analyze real-life situations to the point of recognizing what computations must be made to solve real-life problems has not entered the mainstream of functional citizenship.

It is dangerous to assume that skills from one era will suffice for another. Skills are tools. Their importance rests in the needs of the times. Skills once considered essential become obsolete, and this is likely to increase in pace and scope as advances in technology revolutionize our individual, social, and economic lives. Necessary new skills arise from the dimensions of the mathematics pertinent to an age of population explosion, space exploration, economic and fiscal complexity, and microelectronic wonders. Time and space for including these new skills in the curriculum must be purchased by eliminating the obsolete.

Insisting that students become highly facile in paper-and-pencil computations such as 3841×937 or $72\,509 \div 29.3$ is time-consuming and costly. For most students, much of a full year of instruction in mathematics is spent on the division of whole numbers—a massive investment with increasingly limited productive return. A small fraction of that time is spent on the skills of problem analysis and interpretation, which enable students to identify and set up the computations needed. For most complex problems, using the calculator for rapid and accurate computation makes a far greater contribution to functional competence in daily life.

Common sense should dictate a reasonable balance among mental facility with simple basic computations, paper-and-pencil algorithms for simple problems done easily and rapidly, and the use of a calculator for more complex problems or those where problem analysis is the goal and cumbersome calculating is a limiting distraction.

Reasonable standards of time-effectiveness and cost-effectiveness should be applied to the use of instructional time, where the criterion is the productive applicability of the learned technique to real-life problems.

Professional knowledge of future trends, industrial, financial, engineering, and scientific need, and the demands of daily life are all better arbiters of what is currently essential and what has become obsolete than our nostalgia as parents or teachers.

Recommended Actions

2.1 *The full scope of what is basic should contain at least the ten basic skill areas identified by the National Council of Supervisors of Mathematics's "Position Paper on Basic Skills." These areas are problem solving; ap-*

plying mathematics in everyday situations; alertness to the reasonableness of results; estimation and approximation; appropriate computational skills; geometry; measurement; reading, interpreting, and constructing tables, charts, and graphs; using mathematics to predict; and computer literacy.

2.2 *The identification of basic skills in mathematics is a dynamic process and should be continually updated to reflect new and changing needs.*

2.3 *Changes in the priorities and emphases in the instructional program should be made in order to reflect the expanded concept of basic skills.*

- There should be increased emphasis on such activities as—
    - —locating and processing quantitative information;
    - —collecting data;
    - —organizing and presenting data;
    - —interpreting data;
    - —drawing inferences and predicting from data;
    - —estimating measures;
    - —measuring using appropriate tools;
    - —mentally estimating results of calculations;
    - —calculating with numbers rounded to one or two digits;
    - —using technological aids to calculate;
    - —using ratio and proportion to deal with rate problems in general and with percent problems in particular;
    - —using imagery, maps, sketches, and diagrams as aids to visualizing and conceptualizing a problem;
    - —using concrete representations and puzzles that aid in improving the perception of spatial relationships.
- There should be decreased emphasis on such activities as—
    - —isolated drill with numbers apart from problem contexts;
    - —performing paper-and-pencil calculations with numbers of more than two digits;
    - —mastering highly specialized vocabulary not useful later either in mathematics or in daily living;
    - —converting measures given in one system to corresponding measures in another system;
    - —working with tables whose usefulness as aids to calculation has been supplanted by calculators and other technological aids (e.g., numerical computations with logarithms and cologs).

2.4 *Teachers should incorporate estimation activities into all areas of the program on a regular and sustaining basis, in particular encouraging the* *use of estimating skills to pose and select alternatives and to assess what a reasonable answer may be.*

2.5 *Teachers should provide ample opportunities for students to learn communication skills in mathematics. They should systematically guide students to read mathematics and to talk about it with clarity.*

2.6 *The higher-order mental processes of logical reasoning, information processing, and decision making should be considered basic to the application of mathematics. Mathematics curricula and teachers should set as objectives the development of logical processes, concepts, and language, including—*

- —the identification of likenesses and differences leading to classification;
- —understanding, making, and applying definitions;
- —the development of a feeling for informal proof including counterexamples and generalizations;
- —precise use of such language as *at least, at most, either-or, both-and,* and *if-then.*

Recommendation 3
MATHEMATICS PROGRAMS MUST TAKE FULL ADVANTAGE OF THE POWER OF CALCULATORS AND COMPUTERS AT ALL GRADE LEVELS

Beyond an acquaintance with the role of computers and calculators in society, most students must obtain a working knowledge of how to use them, including the ways in which one communicates with each and commands their services in problem solving.

The availability of computing aids, including computers and calculators, requires a reexamination of the computational skills needed by every citizen. Some of these computational skills will no longer retain their same importance, whereas others will become more important.

It is recognized that a significant portion of instruction in the early grades must be devoted to the direct acquisition of number concepts and skills without the use of calculators. However, when the burden of lengthy computations outweighs the educational contribution of the process, the calculator should become readily available.

With the increasing availability of microcomputers at decreasing costs, it is imperative that schools play an active part in preparing students of the 1980s to live in a world in which more and more functions are being performed by computers.

Recommended Actions

3.1 *All students should have access to calculators and increasingly to computers throughout their school mathematics program.*

- Schools should provide calculators and computers for use in elementary and secondary school classrooms.
- Schools should provide budgets sufficient for calculator and computer maintenance and replacement costs.

3.2 *The use of electronic tools such as calculators and computers should be integrated into the core mathematics curriculum.*

- Calculators should be available for appropriate use in all mathematics classrooms, and instructional objectives should include the ability to determine sensible and appropriate uses.
- Calculators and computers should be used in imaginative ways for exploring, discovering, and developing mathematical concepts and not merely for checking computational values or for drill and practice.
- Teachers should ensure in their classroom management that the use of computers by individual students in isolated activity does not replace the critical classroom interaction of students with peers and teacher. The healthy give-and-take of group work and discussion, which promotes values of communication, cooperation, empathy, mutual respect, and much of cognitive development, remains essential.

3.3 *Curriculum materials that integrate and require the use of the calculator and computer in diverse and imaginative ways should be developed and made available.*

- Schools should insist that materials truly take full advantage of the immense and vastly diverse potential of the new media. In particular, developers of software should be cautioned that just to use conventional material and techniques newly translated to the medium of the computer will not suffice.
- Educators should take care to choose software that fits the goals or objectives of the program and not twist the goals and developmental sequence to fit the technology and available software.

3.4 *A computer literacy course, familiarizing the student with the role and impact of the computer, should be a part of the general education of every student.*

- In cooperation with schools and professional teacher organizations, funding agencies should support the development of courses in computer literacy for both junior and senior high school levels.

3.5 *All mathematics teachers should acquire computer literacy either through preservice programs or through in-service programs funded by school districts in order to deal with the impact of computers on their own lives and to keep pace with the inevitable sophistication their students will achieve.*

- Colleges should provide courses for both perservice and in-service education in computer literacy, programming, and instructional uses of calculators and computers.

- Professional organizations should provide information through their various media, conferences, workshops, and seminars to aid in the in-service education of teachers in uses of the calculator and computer.

3.6 *Secondary school computer courses should be designed to provide the necessary background for advanced work in computer science.*

- Curriculum design should provide the required foundation for those students who will be involved in careers that increasingly demand advanced computing skills and applications of computing and for those students who will go on to deeper study in frontier fields of computer development.

3.7 *School administrators and teachers should initiate interaction with the home to achieve maximum benefit to the student from the coordinated home and school use of computers and calculators.*

- Criteria should be developed to assist parents and school personnel in their selection of home/school computing hardware.

- Professional organizations of teachers, mathematicians, and computer scientists should develop guidelines to aid schools, teachers, and parents in the selection of educational software.

- The uses of technological devices such as calculators, computers, video disks, and electronic games in the home and other out-of-school places should be anticipated. Programs should be planned that will encourage the positive and educationally beneficial use of these devices.

- As home computers come into wider use, homework should be assigned that can take advantage of their potential in problem solving.

3.8 *Educational users of electronic technology should demand a dual responsibility from manufacturers: the development of good software to promote the problem-solving abilities of the student and, eventually, the standardization and compatibility of hardware.*

3.9 *Provisions should be made by educational institutions and agencies to help in the necessary task of educating society's adults in computer literacy and programming.*

3.10 *Teachers of other school subjects in which mathematics is applied should make appropriate use of calculators and computers in their instructional programs.*

3.11 *Teacher education programs for all levels of mathematics should include computer literacy, experience with computer programming, and the study of ways to make the most effective use of computers and calculators in instruction.*

3.12 *Certification standards should include preparation in computer literacy and the instructional uses of calculators and computers.*

Recommendation 4
STRINGENT STANDARDS OF
BOTH EFFECTIVENESS AND EFFICIENCY
MUST BE APPLIED
TO THE TEACHING OF MATHEMATICS

What is learned relative to a topic, how long it is retained, how readily it is applied—all these depend on the learning process the students pass through and how effectively they are engaged in that process. It is fruitless to consider topics taught apart from the way learners meet these topics.

Instructional time is a precious commodity. It must be spent wisely. Learning is a product of both the time engaged in a learning task and the quality of that engagement. Teachers must employ the most effective and efficient techniques at their command. They must apportion instructional time according to the importance of the topic, recognizing that the value of a skill or knowledge is subject to change over time.

Modern technology and educational theory and research have made accessible to today's teacher approaches, materials, and strategies that were not previously available. Teachers at all levels must learn to use this enriched variety of instructional techniques, materials, and resources to teach mathematics more effectively.

Recommended Actions

4.1 *The major emphasis on problem solving in the curriculum must be accommodated by a reprogramming of the use of time in the classroom.*

- Priority in classroom time should be devoted to involving students in meaningful problem-solving activities. Explanation, practice, and directive teaching are important but should not diminish the time necessary to achieve this priority. Requiring complete mastery of skills before allowing participation in challenging problem solving is counterproductive.

211

- The time spent on mathematics in elementary school programs should be increased. Higher-order skills in problem solving require more time to learn than the lower-order, narrowly mechanistic skills.
- The extent of teacher and student time devoted to certain traditional skill areas should be reduced to make room for newly emerging objectives.
- There are certain algorithmic skills (e.g., long division with multiple-digit divisors) that require a great expenditure of classroom time. A strict standard of time-effectiveness and cost-effectiveness should be applied to determine whether actual use of that technique in life outside school justifies this much expenditure of effort and time. The use of calculators has radically reduced the demand for some paper-and-pencil techniques.
- Teachers should learn effective techniques of classroom management to assess and achieve the optimal time on a task.

4.2 *School administrators and parents must support the teacher's efforts to engage students more effectively in learning tasks.*

- Local administrators should ensure uninterrupted time for the teacher to carry out the instructional program.
- Parents and administrators must support the authority of the teacher to require that students be productively engaged in learning during their class time. They should exercise reasonable sanctions against students who do not respect that authority.
- Parents and administrators should expect that teachers will assign a reasonable amount of meaningful homework to extend productively the time students are engaged in the study of mathematics. Both should use their influence to increase the likelihood that students will complete the homework as assigned.
- School budgets should provide for a range of instructional resources adequate to support a wide variety of teaching strategies.

4.3 *Teachers should use diverse instructional strategies, materials, and resources, such as—*

—individual or small-group work as well as large-group work;

—well-planned use of media, such as overhead projectors, videotapes, video disks, audio/video cassettes, computers, films, slides, television;

—the provision of situations that provide discovery and inquiry as well as basic drill;

—the use of manipulatives, where suited, to illustrate or develop a concept or skill;

212

—the inclusion of cyclic review of past topics (contents, skills, and ideas previously taught);

—the use of materials and references outside the classroom, such as visiting museums, using the library, visiting businesses or industries, visiting computer centers, making home television assignments.

4.4 *School districts, local and state or provincial officials, manufacturers, and publishers should take a bolder and more imaginative approach to selecting and producing educational hardware and software in order to provide for a curriculum that emphasizes problem solving.*

• Unnecessarily restrictive conditions set by school districts, state and provincial agencies, outdated ground rules of publishing houses, conservative status quo editorial policies, and the like should give way to greater openness and willingness to address future needs.

Recommendation 5
THE SUCCESS OF MATHEMATICS PROGRAMS AND STUDENT LEARNING MUST BE EVALUATED BY A WIDER RANGE OF MEASURES THAN CONVENTIONAL TESTING

The first purpose of meaningful evaluation in school mathematics should be the improvement of learning programs, teaching, and materials. Educators must evaluate to have information for sound decisions, to be accountable to their public, and to know how well they are doing. Evaluation is a part of mathematics teaching, and hence mathematics educators should be centrally involved in the evaluation process.

Evaluation is not limited to testing. It includes gathering data and interpreting the data. Testing is one source of data. There are many others. Today, many people use test scores as the sole index of the quality of mathematics programs or of the success of student achievement. Test scores alone should not be considered synonymous with achievement or program quality. A serious danger to the education of our youth is the increasing tendency on the part of the public to assume that the sole objective of schooling is a high test score. This is often assumed without the critical knowledge of what is being tested or whether test items fit desired goals.

The evaluation of problem-solving performance will demand new approaches to measuring. Certainly present tests are not adequate. In particular, the measuring of the use of problem-solving processes will demand innovative techniques. Evaluation of programs with problem-solving goals must be sensitive to the nature of those goals.

213

It is imperative that the goals of the mathematics program dictate the nature of the evaluations needed to assess program effectiveness, student learning, teacher performance, or the quality of materials. Too often the reverse is true: the tests dictate the programs, or assumptions of the evaluation plan are inconsistent with the program's goals.

Recommended Actions

5.1 *The evaluation of mathematics learning should include the full range of the program's goals, including skills, problem solving, and problem-solving processes.*

- The evaluation of the use of problem-solving processes must be given special attention by schools, teachers, researchers, test developers, and teacher educators.
- Assessment programs, such as the National Assessment and some state assessments, should continue to sample a wide variety of mathematics learning outcomes and should consider future as well as present needs and programs.
- The development of problem-solving skills should be assessed for each student over the entire K–12 school mathematics program.
- Minimal competencies should not be construed as an adequate measure of an individual's mathematics achievement. What is minimal for all is optimal for none.

5.2 *Parents should be regularly and adequately informed and involved in the evaluation process.*

- With mutual respect (the educator for the sincere concern and valid input of the parent and the parent for the professional expertise of the educator), school administrators, teachers, and parents should cooperate in determining educational goals and the appropriate plan for evaluation.
- Nontest evaluation methods and strategies should be discussed with parents, students, and the general public.

5.3 *Teachers should become knowledgeable about, and proficient in, the use of a wide variety of evaluative techniques.*

- Preservice and in-service teacher education should provide mathematics teachers with knowledge about, and skill in, evaluation.
- Evaluation strategies that include both test and nontest techniques should be developed and disseminated to mathematics teachers both in their initial preparatory programs and in continual in-service updating.
- Teachers should improve their diagnostic skills and their ability to structure appropriate remediation.

- Student strengths and weaknesses should be assessed on a regular basis, using a variety of measures and techniques.
- The informed judgment of teachers should be considered a vital part of the evaluation of any student.
- Teachers should be prepared in the development and use of teacher-made tests.

5.4 *The evaluation of mathematics programs should be based on the program's goals, using evaluation strategies consistent with these goals.*

- Standardized tests should be used in program evaluations only when it can be clearly demonstrated that the test matches the goals of the program.
- The results of a test designed for purposes other than program evaluation (such as the SAT) should not be interpreted as an evaluation of mathematics programs.
- Available evaluation and testing techniques should not determine the goals and objectives of the mathematics program or the emphases of classroom instructional effort.
- Test designers should give attention to the need for more options in format than the conventional multiple-choice format. An emphasis on problem solving demands more flexibility and creativity in assessment than is possible within the restrictions of most current test formats.
- Where minimal competency tests are mandated, they should be implemented with extreme caution to assure that adverse effects on the program do not result.
- Task forces involving parents, teachers, and students should be created to monitor the effects of minimal competency programs.
- Mathematics educators should be centrally involved in the development of competency or assessment programs at local, state, and provincial levels and in the monitoring of the effects of competency testing on mathematics learning.
- Longitudinal evaluation of individual problem-solving ability should be developed. The acquisition of problem-solving skills is a long-term process and should not be evaluated solely with short-term measures.
- Test scores should not be used as the sole index of success in mathematics programs.
- Accreditation of school mathematics programs should use criteria specific to the quality of the mathematics programs rather than to conditions peripheral to content and instructional goals.

5.5 *The evaluation of materials for mathematics teaching should be an essential aspect of program planning.*

- Textbook materials should be judged and selected in terms of the program's goals rather than vice versa.
- Instructional materials with sexist and ethnic biases should not be selected.
- The selection of tests should involve a careful review by teachers as well as administrators.
- Strategies for evaluating nonprint materials must be developed and used.

5.6 *Mathematics teachers must undergo continuing evaluation as a necessary component in improving mathematics programs.*

- Teachers should maintain an awareness of the need and the strategies for self-evaluation.
- A variety of supportive evaluation strategies, such as peer observation, supervisor observation, and videotaping, should be made available to the teacher.
- Any evaluation of mathematics teaching should be sensitive to the instructional goals and should be unique to the content, the teacher, and the class.
- If an evaluation of a teacher's effectiveness includes student performance, the measures of student performance should be consistent with instructional goals. Such evaluation should also consider that there are external factors affecting student performance that are not amenable to teacher influence.
- Judgments of teacher competency are necessary but should be made with caution, with the realization that the validity of most existing measures rests on a shaky foundation.
- Teachers and teacher educators should be centrally involved in the development of instruments for the evaluation of teaching effectiveness.

5.7 *Funding agencies should support research and evaluation of the effects of a problem-solving emphasis in the mathematics curriculum.*

- The nature of problem-solving ability suggests that longitudinal studies will be most meaningful. The more typical short-term project may force a hasty and superficial treatment of programs whose objectives must be complex, interrelated, and of a long-lasting character.

216

Recommendation 6

MORE MATHEMATICS STUDY MUST BE REQUIRED FOR ALL STUDENTS AND A FLEXIBLE CURRICULUM WITH A GREATER RANGE OF OPTIONS SHOULD BE DESIGNED TO ACCOMMODATE THE DIVERSE NEEDS OF THE STUDENT POPULATION

Mathematics is pervasive in today's world. Mathematical competence is vital to every individual's meaningful and productive life. It is, moreover, a valuable societal resource, and the potential of our educated citizenry to make significant use of mathematics is not being fully met. The amount of time the majority of students spend on the study of mathematics in school in no way matches the importance of mathematical understanding to their lives, now and in the future.

Polls show that the public recognizes that mathematics is an essential subject for all students. Yet it is typical to find that only one year of mathematics is required for students in grades 9–12. Surveys of instructional time indicate that only a small percentage of time in elementary school is devoted to mathematics. And only a very tiny portion of that time is spent on teaching children to apply mathematics and solve problems.

When a student discontinues the study of mathematics early in high school, he or she is foreclosing on many options. Many doors, both in college programs and in vocational training, are at once closed to that person. These facts should be communicated more effectively to both the students and their parents.

The recommendation for increasing mathematics study imposes a special responsibility on professional leadership in mathematics education and the schools. To say that most students should study more mathematics is not to say that it should be the same mathematics for all. It does not mean simply keeping all students longer in the same traditional track. In fact, such a recommendation poses a tremendous challenge to curriculum developers and school districts to devise a more flexible range of options, a diversified program to meet a variety of interests, abilities, and goals.

Except for a distinction among courses typically taken in the ninth grade, differentiation in students' mathematics programs has been based mainly on whether or not a student takes elective years of mathematics. If everyone now needs more mathematics study, then differentiation must occur within the program.

The growing diversification of the applications of mathematics in an ever-increasing variety of college programs of study demands more than a single college-preparatory program. At the college level, the trend is to broaden the conception of mathematical study to include what is called "the mathematical sciences." Roughly, this means not only mathematics but the delivery of

217

mathematical ideas and tools to the solution of real-life problems. Present high school programs do not fully anticipate the many options provided by mathematics, the mathematical sciences, and computer science.

Technical and vocational training at different levels also assumes more and diverse mathematical backgrounds. For those whose formal education will end with high school, the needs of citizen and consumer for increasing mathematical sophistication dictate a collection of courses based on consumer and career needs, computer literacy, and quantitative literacy.

It is important that recommended programs permit lateral movement and not strictly "track" students, trapping them in a linear pattern that does not permit change to another path. Flexibility is vital, and the key is to keep options open as long as possible.

Since a higher level of mathematical skill and understanding will increasingly become a significant advantage in nearly all lives, then justice demands that all groups have equal access to these advantages. At present, females and some minority groups are underrepresented in mathematics courses and courses for which mathematics is prerequisite.

It is naive to suppose that just providing mathematics courses as electives will serve for equality of opportunity. All reasonable means should be employed to assure that everyone will have the foundation of mathematical learning essential to fulfilling his or her potential as a productive citizen. The currently underrepresented groups should be especially encouraged and helped.

Recognizing diversified individual interests and needs entails devising programs that are tailored for particular categories of students. Differentiated curricula must incorporate the special needs in mathematics of students with handicaps, including physical or learning difficulties. These programs will need to move away from the idea that everyone must learn the same mathematics and develop the same skills. Mathematics and mathematical ability cover a much broader range than most people realize.

In many current programs, the student who does poorly in the algorithmic skills finds progress in all aspects of mathematical development halted, since remediation is designed to concentrate solely on this deficiency. Remedial programs should identify other areas of mathematical ability—for example, spatially related skills—and concentrate attention also on the students' strengths, not solely on their deficits.

The student most neglected, in terms of realizing full potential, is the gifted student of mathematics. Outstanding mathematical ability is a precious societal resource, sorely needed to maintain leadership in a technological world.

Mathematics educators and curriculum developers should redesign the sequence of the curriculum to realize the critical process goals of problem solving, as well as content and skill goals. A clear and logical developmental sequence for process objectives from kindergarten through twelfth grade should be described and serve as an organizing framework.

In very general terms, such a sequence should proceed through stages of development, though in practice, progression should be smooth and unbroken.

In the elementary school, children, in their experiences in solving particular problems, gradually develop their higher-order mental abilities and learn particular skills and particular strategies. At this stage, strategies are primarily specific to individual problems. A foundation of skills must be laid, but skills should not be learned entirely in isolation from application even in the primary grades. There should be an interplay of skill building and application throughout.

Moving through the upper grades and into the junior high school level, the progress should be toward more generalization, more abstraction of techniques, more emphasis on similarities and patterns found in differing contexts. Strategies become not just ways to solve one type of problem but generalized and synthesized. Techniques learned in one context may be recognized as applicable to other problems. Specific attention is needed to help students make the important transition to the abstract reasoning processes.

During the seventh and eighth grades, intensive focus on problem solving should become a vehicle to exercise, confirm, and develop further all basic skills. At the same time, familiarity, competence, and confidence should be built in applying these mathematical skills to solving problems of varying difficulty and from diverse settings. At this stage, a significant skill is the ability to select strategies from a growing repertoire.

A broad range of problem-solving approaches should be explored so that the teacher can identify students' special strengths and advise them on high school options.

The ability to create strategies to attack a new problem is, at the junior high school level, simple and embryonic. It should increase in sophistication throughout high school. More flexibility and power should be achieved by more generalization and abstraction. The essential nature of techniques and strategies and their range of applicability should be emphasized as students begin to see more applications to broad categories of problem areas, as in, for example, application to other disciplines. There should be a greater effort made to coordinate the mathematics learned in mathematics classes with other subjects that use mathematics. Students who perform well in mathematics classes often fail to see any relationship in what they are doing there to the mathematical techniques employed in classes in science and other subjects.

This does *not* suggest that the content of high school mathematics is either dictated by or limited to the range of mathematical techniques used in other subjects. The goal is to develop a more flexible, deeper, and broader problem-solving power, and this goes significantly beyond the formulas and recipes that have been traditionally applied to familiar problems.

At the same time, the important interplay and integration of mathematics and its applications in learning should not cease because isolated course structure separates mathematics from disciplines that apply it. The principles of learning must take priority over administrative convenience. It is likely that this coordination will have to be accomplished by voluntary cooperative efforts on the part of mathematics teachers and the teachers of other subjects that apply mathematics, with a healthy mutual respect for the legitimate goals of both groups.

At the high school level many students can apply their problem-solving abilities not only to problems of daily life, to problems from other disciplines, but to serious mathematical problems themselves.

Recommended Actions

6.1 *School districts should increase the amount of time students spend in the study of mathematics.*

- At least three years of mathematics should be required in grades 9 through 12.
- The amount of time allocated to learning mathematics in elementary school should be increased. It should range from a minimum of about five hours a week in the primary grades to a minimum of about seven hours a week in the upper grades. Part of this time can be gained by a program that stresses the application of mathematical skills in other subjects, especially the sciences and the social studies.
- A clear delineation of what constitutes college-level and precollege mathematics should be made.
- Colleges should not award college credit for courses in which the level of content is that of high school mathematics. This practice encourages students not to elect mathematics in high school beyond the minimum required.

6.2 *In secondary school, the curriculum should become more flexible, permitting a greater number of options for a diversified student population.*

- Increasing high school requirements in mathematics should not result in keeping all students longer in the same traditional tracks. These recommendations cannot be met with just the two-track alternatives of either general mathematics courses or precalculus courses typical of many existing programs.
- The high school curriculum should provide differing student populations with those appropriately organized areas of mathematical competence required by their needs, talents, and future objectives, but all presented with continual attention to functional problem-solving ability.
- Algebra should be included in the program of all capable students to keep their options open.
- For many students, algebra should be delayed until a level of maturity and basic mathematical understanding permit their taking full advantage of a significant algebra course. For many, this may not be ninth grade but perhaps eleventh or even twelfth grade. Significant

220

mathematics courses should be available to these students in ninth and tenth grades, not just the traditional general mathematics review or prealgebra courses.

- If such recommendations are followed, a course providing progress beyond junior high school but paving the way for a useful experience in consumer mathematics and later algebra needs to be developed.

- Consumer mathematics should develop a broader quantitative literacy and should consist primarily of work in informal statistics, such as organizing and interpreting quantitative information.

- All high school students should have work in computer literacy, and the hands-on use of computers, and the applications of computers where possible and appropriate throughout their mathematics programs.

- All students who plan to continue their study of mathematics beyond high school or to use it extensively in technical work or training should be enrolled in mathematics courses throughout their last high school year.

6.3 *Mathematics educators and college mathematicians should reevaluate the role of calculus in the differentiated mathematics programs.*

- Emerging programs that prepare users of mathematics in nontraditional areas of application may no longer demand the centrality of calculus that has traditionally been demanded for all students. (The Mathematical Association of America's PRIME 80 conference raised questions about the role of calculus as the eventual touchstone that dictated all college preparatory mathematics.)

- In light of this reevaluation, colleges and high schools should reexamine the concept of advanced placement in mathematics. For some students, though fully capable, an advanced placement program restricted to calculus placement may not be the optimal alternative. If advanced placement in mathematics is encouraged, it should be a broader concept that includes options in other branches of the mathematical sciences.

6.4 *The curriculum that stresses problem solving must pay special heed to the developmental sequence best suited to achieving* process *goals, not just* content goals.

- From the earliest years, the basic mathematical tools should be acquired within the framework of usage and application, however simple the examples may be at the beginning levels.

- Since there are usually multiple approaches to all but the most trivial problems, not all students should be expected to proceed in the same way. Value should be placed on a thoughtful and productive approach, not solely on a single correct answer.

221

- Since elementary school children differ widely in maturation and intellectual development, the teacher should be prepared to value and reward different contributions made by different students to the solution of a common problem.
- Team efforts in problem solving should be commonplace in the elementary school classroom.
- At middle school and junior high school levels, the focus of the curriculum should be on more formal and more general problem-solving approaches and strategies themselves.
- At middle school and junior high school, instruction should stress the ability to apply techniques used in one situation to new and unfamiliar situations.
- At middle school and junior high school, instruction should stress the ability to select from a range of strategies and to create new strategies by combining known techniques.
- At middle school and junior high school, instruction should aid in the student's transition to more abstract reasoning.
- Difficulty with paper-and-pencil computational algorithms should not interfere with the learning of problem-solving strategies.
- At the junior high school level, calculators should be available so that no student will be excluded from the opportunity to develop these strategies.
- All courses in high school mathematics should include some activities in applications.
- Teachers of mathematics and teachers of other disciplines should cooperate in assuring that students perceive the relationship of the mathematics they learn to the mathematics applied in problems in those other disciplines.
- Qualified mathematics teachers should be used as resource specialists for instructional programs in which mathematics methods are applied in other subjects.
- Teachers of mathematics should be prepared in the application of known problem-solving techniques to a variety of problems.

6.5 *Teachers, school officials, counselors, and parents should encourage a positive attitude toward mathematics and its value to the individual learner.*

- A curriculum that focuses on problem solving should challenge all students. It should present the opportunity for students of all ability levels to make a contribution, and it should promote the attitude that they are capable of solving problems.
- Programs that will encourage a larger percentage of females and

minority students to study more mathematics should be designed and supported.

- Parents and counselors should help students to recognize the importance of mathematics study to their futures and guide them to make appropriate school decisions.

6.6 *Special programs stressing problem-solving skills should be devised for special categories of students.*

- The professionals in mathematics and in special education should work together to identify the process abilities, possible and optimal, for students with handicaps and learning difficulties.
- Perception of what mathematical ability encompasses should be broadened beyond the linear, algorithmic stereotype. The significance of spatial perception and spatial relationships in problem solving needs to be stressed, for many students who do not fare well in algorithmic thinking may have special abilities that are spatial and geometric.
- Increased attention should be paid to developing the potential of the gifted student of mathematics.
- Colleges and schools should cooperate in devising imaginative programs for the mathematically gifted.
- In general, programs for the gifted should be based on a sequential program of enrichment through more ingenious problem-solving opportunities rather than through acceleration alone.
- Materials and resources of a sophistication and depth suitable to the unusual potential of the gifted student in problem solving should be developed.

Recommendation 7
MATHEMATICS TEACHERS MUST DEMAND OF THEMSELVES AND THEIR COLLEAGUES A HIGH LEVEL OF PROFESSIONALISM

Educators can gain the support of society and the rewards of a truly effective performance by developing, defining, and enforcing professional standards in terms of highly competent professional performance rather than by any other norm. This must be done to provide the nation, its young people, and its future with the mathematics programs worthy of them and of that future.

Within mathematics teaching—

—there are already many well-prepared and effective teachers who provide outstanding professional leadership;

223

—there are also many teachers who are motivated and desire to improve but who lack adequate preparation and must be given the necessary support to become fully qualified and to improve;

—there are, however, some teachers whose attitudes and functioning are at less than a professional level. In the best interests of students and society, the number of such teachers must be reduced immediately.

Even the best prepared, competent, and dedicated teachers must continue their development to keep abreast of changing needs, tools, and conditions. School administrations have the responsibility to make this possible by providing continuing in-service education and by encouraging teachers to take full advantage of opportunities for maintaining their competence that are offered by professional organizations and universities.

At the beginning of this decade, the schools are faced with a widespread shortage of qualified mathematics teachers. The demand for mathematical competence in many sectors of society is great and growing, and schools find it impossible to compete for individuals who have this desired background. Thus, in many mathematics classrooms, the teacher does not have the subject-matter qualifications for teaching mathematics. School administrations have two obligations: (1) to be forthright and open with the parents about the situation, and (2) to provide special aid and support needed by these teachers until they can make up deficiencies. The professional organizations also have a special responsibility to cooperate in assisting such teachers who are dedicated to improvement. However, the public and its representatives must give high priority to finding ways to solve the worsening problem.

Regardless of preparation, the standard of professionalism should be consistently high, and it is the obligation of any group that wishes to be called a profession to insist that all members maintain this standard.

Teachers must be sensitive to the needs of their students and dedicate themselves to the improvement of student learning as their primary professional objective. The right of students and parents to expect this dedication has correlative responsibilities: a mutual respect and support by the parent of the educational program and a teacher's professional competence, and the acceptance by the student of ultimate personal and active commitment to his or her own learning.

Any teacher who lacks dedication to these professional ideals and to continued self-improvement should not be retained in teaching. Teachers must accept performance and not protectionism as a synonym for professionalism.

During the decade of the 1980s, the continuing appearance of new concepts and theories in mathematics, in the applications of mathematics, and in the teaching-learning process will affect both curriculum and instruction in school mathematics. In order to remain professional, teachers must continue to study in all three areas. This will require of teachers a new level of motivation and dedication.

Recommended Actions

7.1 *Every mathematics teacher should accept responsibility for maintaining teaching competence.*

- Full advantage should be taken of all existing opportunities for continuing education.
- Teachers should insist that school districts and colleges make provision for in-service education and staff development opportunities.
- Teachers should belong to professional organizations that are dedicated to the improvement of teaching and learning.
- Teachers should participate actively in the efforts of professional organizations to improve teaching and learning.
- Teachers should share ideas and participate with their peers in cooperative efforts at self-improvement, including observation and constructive criticism of one another.
- Teachers as a profession should insist that all members maintain a consistently high standard of professional behavior. The profession is not obligated to protect those individuals who refuse to live up to reasonable professional standards.
- Bargaining units of organizations representing teachers should include in their requests release time for professional development and attendance at professional conferences that provide in-service education.

7.2 *School boards and school administrations should take all possible means to assure that mathematics programs are staffed by qualified, competent teachers who remain current in their field.*

- Necessary incentives must be found to attract competent and dedicated teachers to the profession.
- The status, compensation, and teaching conditions necessary for the retention of qualified teachers must be dramatically improved.
- School districts must budget adequately and provide incentives for teachers to participate in in-service education pertinent to their immediate needs as they prepare to meet the challenges of the future.
- School district staffs, including teachers, must plan in-service education that is articulated with local colleges and universities as well as with professional organizations such as the NCTM and its state and local affiliates.
- School administrators should encourage teachers to take an active professional role and should permit them to participate, without penalty, in conferences and vital professional work.
- School systems should maintain well-qualified mathematics specialists or supervisors at all levels to help teachers achieve the pro-

225

fessional level specified in these recommendations and to coordinate mathematics education efforts within the system.

- Well-qualified mathematics specialists should be on the staffs of all governmental agencies that deal with mathematics education—specifically, in all state and provincial departments of education, in the National Education Department, the National Institute of Education, and the National Science Foundation.

7.3 *Teacher education institutions and agencies should develop new programs of preparation to incorporate the problem-solving emphases recommended.*

- Programs in teacher education must be designed to prepare teachers for new levels of performance and professionalism.
- The effective teaching of problem solving requires thorough preparation both in mathematical content and in teaching methods that develop problem-solving ability.
- Colleges and universities must redesign courses and programs to incorporate revisions recommended for the preparation of mathematics teachers.

7.4 *Certification standards for mathematics teaching should be revised and upgraded to incorporate the needs reflected in these recommendations.*

- Professional organizations should participate in defining standards for certification of teachers.
- State and provincial departments of education should involve professional organizations in setting professional standards and defining qualifications.
- Professional organizations—such as the Mathematical Association of America, Conference Board of the Mathematical Sciences, National Council of Supervisors of Mathematics—in cooperation with the NCTM should continually review and update their guidelines for the preparation of teachers of mathematics.

Recommendation 8

PUBLIC SUPPORT FOR MATHEMATICS INSTRUCTION MUST BE RAISED TO A LEVEL COMMENSURATE WITH THE IMPORTANCE OF MATHEMATICAL UNDERSTANDING TO INDIVIDUALS AND SOCIETY

Solutions to the problems identified in previous sections cannot be achieved solely within the education community but require active participation and support by parental and societal groups.

The mathematics teaching profession recognizes and respects the right of parents and society to hold it accountable for the mathematical competence of children. However, in calling for particular programs of action, parents and society often mistakenly promote activities that are counterproductive to the realization of the goals they support. Communication and cooperation must bridge this gap.

The rededication of teachers called for in Recommendation 7 will be meaningless or impossible if society is not committed to supporting the professionalism it rightfully demands. Today, teachers' financial incentives are neither commensurate with the responsibilities they carry nor adequate to attract and retain them in a demanding, crucial, and sometimes burdensome profession. Few school systems have truly adequate supervisory and material support for the teacher in the maintenance and improvement of the instructional environment. More and more, teachers feel the lack of parental understanding of the complexity of their task and the lack of parental cooperation and support in their efforts to instruct children. Furthermore, governmental support for improving the quality of mathematics teaching has dwindled.

Essential to the success of the program outlined in these recommendations are (1) the willingness of government and private funding agencies to listen and respond more sensitively to the professionals in elementary and secondary mathematics education, and (2) the reestablishment by such agencies of institutes and other programs for the continuing education and upgrading of teachers.

At present, there are too many unnecessary obstacles to the effective functioning of teacher and student in a true teaching/learning interaction. These include more and more time required for unproductive record keeping; many unmotivated, undisciplined students; a lack of parental support; ambivalence and vacillation in government regulations; shifting societal priorities; the lack of home and school agreement on out-of-school study assignments.

There are also obstacles faced by school administrators and by the policy-making bodies, the school boards, whose ultimate responsibility is the effectiveness of the school district. It is recognized that in many instances confusions of legalisms and court decisions impose constraints on the ability of these people to accomplish what they agree should be done. Society today is troubled with many complexities. All institutions are threatened. These obstacles should not be an excuse for inaction; rather, in the light of pressing student needs it is imperative that they be surmounted.

All these things make the ideal professional role of the teacher almost impossible; they make it difficult to attract and retain teachers of a level of competence necessary to realize the outlined program. The immediate loser is the individual student, who is excluded from full participation in, and contribution to, society. The ultimate loser is society itself.

The public and its representatives need to confront a serious problem that is increasing in magnitude. There are not enough qualified mathematics teachers to fill mathematics classroom positions. Present enrollments in teacher preparation programs guarantee that the shortage will increase. Mathematics is of critical importance and should be taught by those well qualified and knowledgeable. Such knowledge is not gained in a crash program of short duration. Parents concerned for their children's future and society concerned for its own future should immediately find extraordinary measures to solve this problem.

The professional community and society share a common goal: to bring all citizens to the full realization of their mathematical capacity. This is a complex and delicate task, and it requires the commitment and cooperation of all segments of society, not just the school, parents, and teachers.

Recommended Actions

8.1 *Society must provide the incentives that will attract and retain competent, fully prepared, qualified mathematics teachers.*

- School districts must provide compensation commensurate with the professionalism and qualifications necessary to achieve our educational goals in mathematics.

- School districts should investigate a variety of incentives and conditions that will stop the drain of qualified mathematics teachers to other, more highly compensated fields of work.

- School districts should assure mathematics teachers a classroom environment and conditions conducive to effective teaching, including a reasonable class size.

- School districts should provide teaching conditions and incentives that will attract dedicated and competent people into the mathematics teaching profession.

8.2 *Parents, teachers, and school administrations must establish new and higher standards of cooperation and teamwork toward the common goal of educating each student to his or her highest potential.*

- Professionals must respond to calls for maintaining educational standards and must work cooperatively with parents to specify these standards.

- Parents must support the maintenance of agreed on standards of achievement and discipline.

- Programs in mathematics that take full advantage of home and school cooperation should be systematically developed.

- Parents should enter the process of determining educational goals as partners, with shared participation and responsibility for the accomplishment of those goals.

- Parents should support the teacher's assignment of homework when it is reasonable and clearly related to the educational objectives.

228

- Parents should help guide students to an understanding of their critical need to learn mathematics.
- Parents should help students understand that they have the ultimate responsibility for their own learning and must be active and cooperative participants in the process.
- Parents and teachers should cooperate in a mutually supportive attack on the erosion of respect for authority in the classroom.
- Parents and teachers should cooperate in a mutually supportive attack on the erosion of motivation toward academic achievement.

8.3 *Government at all levels should operate to facilitate, not dictate, the attainment of goals agreed on cooperatively by the public's representatives and the professionals.*

- Government funding agencies should support an emphasis on research and development in applying mathematics to problem solving.
- Teachers should have advisory roles in all decisions of policy and support.
- When legislation affecting education is required, it should be entered into with caution and only after the involvement of educational professionals in formulating and reviewing the mathematical and pedagogical aspects of such legislation.
- Legislation concerning accountability should take into account the multiple factors that determine school achievement.
- Legislators should avoid tendencies to mandate testing as the sole criterion for the evaluation of educational success.
- Mandates for the achievement of minimal competencies should not limit the school mathematics program in its broader range of essential goals.
- Legislation should not determine educational goals but when necessary should facilitate the achievement of cooperatively agreed on objectives.
- Dual respect and effective articulation must become commonplace between the civic leaders who appropriately call for educational accountability and the educational professionals who must formulate responses.